Security Informatics and Law Enforcement

Series Editor
Babak Akhgar, CENTRIC
Sheffield Hallam University, Sheffield, UK

Editor's Note:

The primary objective of this book series is to explore contemporary issues related to law enforcement agencies, security services and industries dealing with security related challenges (e.g., government organizations, financial sector insurance companies and internet service providers) from an engineering and computer science perspective. Each book in the series provides a handbook style practical guide to one of the following security challenges:

Cyber Crime - Focuses on new and evolving forms of crimes. Books describe the current status of cybercrime and cyber terrorism developments, security requirements and practices.

Big Data Analytics, Situational Awareness and OSINT- Provides unique insight for computer scientists as well as practitioners in security and policing domains on big data possibilities and challenges for the security domain, current and best practices as well as recommendations.

Serious Games – Provides an introduction into the use of serious games for training in the security domain, including advise for designers/programmers, trainers and strategic decision makers.

Social Media in Crisis Management – explores how social media enables citizens to empower themselves during a crisis, from terrorism, public disorder, and natural disasters

Law enforcement, Counterterrorism, and Anti-Trafficking – Presents tools from those designing the computing and engineering techniques, architecture or policies related to applications confronting radicalisation, terrorism, and trafficking.

The books pertain to engineers working in law enforcement and researchers who are researching on capabilities of LEAs, though the series is truly multidisciplinary – each book will have hard core computer science, application of ICT in security and security / policing domain chapters. The books strike a balance between theory and practice.

More information about this series at
http://www.springer.com/series/15902

Babak Akhgar • Marco Gercke
Stefanos Vrochidis • Helen Gibson
Editors

Dark Web Investigation

Springer

Editors
Babak Akhgar
CENTRIC
Sheffield Hallam University
Sheffield, UK

Marco Gercke
Cybercrime Research Institute
Köln, Nordrhein-Westfalen, Germany

Stefanos Vrochidis
Information Technologies Institute
Centre for Research and
Technology Hellas
Thessaloniki, Greece

Helen Gibson
CENTRIC
Sheffield Hallam University
Sheffield, UK

ISSN 2523-8507 ISSN 2523-8515 (electronic)
Security Informatics and Law Enforcement
ISBN 978-3-030-55342-5 ISBN 978-3-030-55343-2 (eBook)
https://doi.org/10.1007/978-3-030-55343-2

This Springer imprint is published by the registered company Springer Nature Switzerland AG
The registered company address is: Gewerbestrasse 11, 6330 Cham, Switzerland

Dedicated to Aryan H. Akhgar
You are always in our hearts

Foreword

Criminality on or enabled by the dark web has been a concern for law enforcement agencies (LEAs) since before the Silk Road dark web market came to mainstream prominence in 2011. Many years later, the difficulties faced in policing the dark web remain a key challenge for law enforcement.

The issues encountered by LEAs in conducting investigations on the dark web are wide and varied including technical and ethical challenges such as network access and configurations, limited access to tools which support dark web investigation (e.g. even simple access to the ToR browser), a mindset that is limited to manual investigatory processes, budgetary concerns that prevent the purchase of specialised software to support investigations, and lack of knowledge of both investigators and senior investigators in terms of 'how' to access and investigate content on the dark web. Even the name, 'the dark web' conjures up a form of fear and mystique that leads less experienced investigators to believe that accessing the dark web is somehow illegal or may cause them to be immediately confronted with overwhelming amounts of terrorist content, child sexual abuse material or the opportunity to buy copious amounts of drugs and weapons.

As criminals become more technically savvy and move more of their operations onto the dark web, LEAs must try to get ahead of them and develop capabilities to prevent and detect the criminal activities. Even as investigatory procedures and technological capabilities improve (as we have seen for open source intelligence (OSINT) in recent years), the challenge remains to have access to the 'right' tools to effectively and lawfully carry out dark web investigations whilst integrating such tools with already fragmented software ecosystems used by LEAs. Such systems usually suffer from a lack of a harmonised data model, duplication of data (especially nominal information), and an over-reliance on either free text entries or screenshots which, while useful against individual pieces of intelligence, restrict the ability of investigators to identify links between persons, locations, conversations and other data items as their investigation grows. Furthermore, the multi-source nature of internet investigations and, by extension, dark web investigations mean they need tools which can automatically gather intelligence from a range disparate and often ephemeral web pages, profiles, forums and markets in heterogeneous formats and combine them with extensive and purposeful processing to fuse such data into a common and comparable representation. Such tools will enable LEAs to safely and lawfully capture the 'what' of what is happening on the dark web and support them to build a picture of the 'how' and the 'why', essential for transitioning from the identification of criminal activity to a conviction.

In online research, LEAs face another final challenge – navigating the web lawfully, ethically, in accordance with the GDPR and the law enforcement directive and doing-so in a manner that is societally acceptable. Tacking such challenges is not a simple process and must be considered from the investigator and user right through to the tool developers and designers.

In these 13 chapters, our aim is to inform, educate and promote discussion on the challenges of performing investigations on the dark web. It is our hope that the first chapters will provide a solid basis from which investigators of all levels of expertise with the dark web can learn and consolidate their understanding. That the next chapters will highlight areas where potential criminal activity on the dark web already takes place and how the development and use of well-designed technological tools can greatly assist in investigations from data collection to analysis. Tackling the ethical, legal and privacy-based challenges are essential for any investigator and here we provide an informed discussion of the many complexities to be considered. Finally, successful dark web investigators can learn from those which have gone before them, case studies of dark web investigations across a range of topics and modalities show how the dark web is just one part of the criminal ecosystem and that disruption and informed enquiries can have a significant impact on the criminal enterprise.

I hope this book is able to support investigators, practitioners and researchers alike, especially those beginning their dark web journey, in enhancing their confidence and understanding of the dark web and how it may provide crucial and underexplored avenues for intelligence gathering within their investigations.

John D. Parkinson, OBE, MSt, CMgr, CCMI, CIMS, FRSA, United Kingdom

Acknowledgements

The editors would like to thank the experts who have contributed to this book, sharing their knowledge, experience and latest research. In particular, we would like to thank the following people, projects and organisations:

- Police Services of Northern Ireland (PSNI)
- European Commission's Horizon 2020 Research and Innovation Programme, the TENSOR project and its partners
- European Commission (EC)
- DG HOME (European Commission)
- Staff at CENTRIC (Centre of Excellence in Terrorism, Resilience, Intelligence and Organised Crime Research)
- Staff at Information Technologies Institute, Centre for Research and Technology Hellas
- The Center for Security Studies (KEMEA)
- Dr Nada Milisavljevic of DG HOME
- Jonathan Middleton of Police Services of Northern Ireland (PSNI)
- Ulrich Gasper of Cybercrime Research Institute
- Mary James of Springer

Contents

Editor Biographies

Babak Akhgar is Professor of Informatics and Director of CENTRIC (Centre of Excellence in Terrorism, Resilience, Intelligence and Organised Crime Research) at Sheffield Hallam University, UK, and Fellow of the British Computer Society. He has more than 130 refereed publications in international journals and conferences on strategic information systems with specific focus on knowledge management (KM) and intelligence management. He is member of editorial boards of several international journals and has acted as Chair and Program Committee Member for numerous international conferences. He has extensive and hands-on experience in the development, management and execution of KM-projects and large international security initiatives (e.g. the application of social media in crisis management, intelligence-based combating of terrorism and organised crime, gun crime, cyber-crime and cyber terrorism, and cross-cultural ideology polarisation). In addition to this, he acts as technical lead in EU security projects (e.g. the EU H2020-project TENSOR on dark web). He has co-edited numerous books on Intelligence Management, for instance, *Emerging Trends in ICT Security*, *Application of Big Data for National Security*, *Open Source Intelligence Investigation* and *Combatting Cybercrime and Cyberterrorism*. Prof Akhgar is board member of the European Organisation for Security (EOS) and member of the Academic Advisory Board of SAS UK.

Marco Gercke is an entrepreneur, thinker and scientist. With more than 1000 speeches in over 100 countries and over 100 scientific publications, Marco is a leading experts in the field of cybersecurity and cybercrime. He is the Founder and Director of the Cybercrime Research Institute, an independent research institute. He advises governments, international organisations and large enterprises around the world with regard to strategic, political and legal issues in the field of cybersecurity. Over the past 15 years, he has worked in over 100 countries across Europe, Asia, Africa, the Pacific and Latin America. Marco is involved in various international projects related to cybersecurity including the EU-funded projects TENSOR and FORESIGHT.

Helen Gibson is a Senior Research Fellow and Operations Lead within CENTRIC at Sheffield Hallam University. Prior to CENTRIC, Helen studied for her PhD at Northumbria University in Graph and Network Visualisation. Before that, she completed a Master's in Computing also at Northumbria and BSc in Mathematics at Edinburgh University. Helen's main research interests lie in the areas of data science and visualisation with a specific focus on how data can be used and presented to achieve the maximum value and understanding in intelligence operations. Helen has worked on a number of EU-funded projects within CENTRIC including Athena, Unity, TENSOR, ROBORDER and CONNEXIONs as well as supporting the development of the open source intelligence capability with CENTRIC.

Stefanos Vrochidis has a Diploma in Electrical Engineering from Aristotle University of Thessaloniki, an MSc degree in Radio Frequency Communication Systems from University of Southampton, and a PhD degree in Interactive Video Retrieval based on implicit user feedback from Queen Mary, University of London. He is a Senior Researcher with the Multimedia Knowledge and Social Media Analytics Lab at the Information Technologies Institute of the Centre for Research and Technology Hellas (CERTH-ITI). Dr. Vrochidis is also a Research Development Manager and Co-founder of Infalia PC. His research interests include multimedia analysis, computer vision, web data mining, semantics, information retrieval, multimodal analytics, decision support, as well as security applications including crisis management, fighting crime and terrorism, and border surveillance.

Dr. Vrochidis has participated in more than **30** national and European projects relevant to ICT and Security, of which he has been the Project Coordinator in three, Deputy Project Coordinator in two, and Scientific/Technical Manager in four. He has been the organiser of various workshops relevant to multimodal retrieval, multimedia, and security and has served as regular reviewer in several scientific journals and conferences. He is also the co-author of more than **180** conference, journal and book chapter **articles.**

Contributors

Babak Akhgar CENTRIC, Sheffield Hallam University, Sheffield, UK

Eleni Darra Center for Security Studies-KEMEA, Athens, Greece

Tony Day CENTRIC, Sheffield Hallam University, Sheffield, UK

Kieran Dennis CENTRIC, Sheffield Hallam University, Sheffield, UK

Ulrich Gasper Cybercrime Research Institute, Cologne, Germany

Marco Gercke Cybercrime Research Institute, Cologne, Germany

Georgios Giataganas Center for Security Studies-KEMEA, Athens, Greece

Helen Gibson CENTRIC, Sheffield Hallam University, Sheffield, UK

Dimitrios Kavallieros Center for Security Studies-KEMEA, Athens, Greece

University of Peloponnese-Department of Informatics and Telecommunications, Tripoli, Greece

Emmanouil Kermitsis Center for Security Studies-KEMEA, Athens, Greece

Euthimios Lissaris Center for Security Studies-KEMEA, Athens, Greece

Dimitrios Myttas Center for Security Studies-KEMEA, Athens, Greece

Alice Raven CENTRIC, Sheffield Hallam University, Sheffield, UK

Yara Abdel Samad CENTRIC, Sheffield Hallam University, Sheffield, UK

PART I

Foundations

CHAPTER 1

Understanding the Dark Web

Dimitrios Kavallieros, Dimitrios Myttas, Emmanouil Kermitsis, Euthimios Lissaris, Georgios Giataganas, and Eleni Darra

1.1 Introduction

Dimitris Avramopoulos, European Commissioner for Migration, Home Affairs and Citizenship, said:

> The Dark Web is growing into a haven of rampant criminality. This is a threat to our societies and our economies that we can only face together, on a global scale…

D. Kavallieros (✉)
Center for Security Studies-KEMEA, Athens, Greece

University of Peloponnese-Department of Informatics and Telecommunications, Tripoli, Greece
e-mail: d.kavallieros@kemea-research.gr

D. Myttas • E. Kermitsis • E. Lissaris • G. Giataganas • E. Darra
Center for Security Studies-KEMEA, Athens, Greece
e-mail: d.myttas@kemea-research.gr; e.kermitsis@kemea-research.gr; e.lissaris@kemea-research.gr; g.giataganas@kemea-research.gr; e.darra@kemea-research.gr

B. Akhgar et al. (eds.), *Dark Web Investigation*, Security Informatics and Law Enforcement,
https://doi.org/10.1007/978-3-030-55343-2_1

Having in mind a number of similar statements around the world and before diving into the Dark Web, it is essential to make an introduction to the principal terms of the "digital world" as it was presented for the first time in the 1960s. Initially, it must be outlined that even though many people use the interchangeable terms Internet and World Wide Web (web), the two terms are not synonymous. The Internet and the web are two separate, but of course, related things.

The Internet is a massive network of networks, and it connects millions of computers together globally, forming a superset in which any computer can communicate with any other computer as long as they are both connected to the Internet. Information that travels over the Internet does so via a variety of "languages" known as protocols (Internet protocol, IP) and through satellite, telephone lines and optical cables forming the global electronic community. The Internet has no centralised governance in either technological implementation or policies for access and usage; each constituent network sets its own policies (Strickland 2014). On the contrary, the World Wide Web, or simply web, is a way of accessing information over the medium of the Internet. It is an information-sharing model that is built on top of the Internet. The web uses the HTTP protocol, only one of the languages spoken over the Internet, to transmit data. Web services, which use HTTP to allow applications to communicate in order to exchange business logic, use the web to share information, consisting of HTML text, images, audio, video and other forms of media. The English scientist Tim Berners-Lee invented the World Wide Web in 1989, as he wrote the first web browser computer program in 1990, and has been employed at CERN in Switzerland. The web browser was released outside CERN in 1991, first to other research institutions starting in January 1991 and to the general public on the Internet in August 1991.

Having in mind that the two terms are not synonymous and should not be confused, easily one can say that the web is just a portion of the Internet, although a large portion of it. Content on the World Wide Web can be broken down into two basic categories: structured and unstructured, while the web consists of several layers of accessibility. The first layer is called the clear web or surface web. Surface web is the portion of the web that is readily available to the general public and searchable with standard web search engines. This part is accessible through regular search engines and is where social media platforms reside also. The surface web has been part of the World Wide Web since the first browser was invented, connecting users with websites that can be discovered through a regular Internet browser (e.g. Edge, Mozilla, Opera, etc.) using any of the main search engines (Google, Yahoo, etc.). This is what you use when you read the news; buy something, e.g. on Amazon; or visit any of your usual daily websites and is also the area of the web that is under constant surveillance by governments across the world. Surface web is made up of static and fixed pages. Static pages do not depend on a database for their content. They reside on a server waiting to be retrieved and are basically html files whose content never changes. Thus, any reference to surface web will be referring to common websites, that is, sites whose domains end in .com, .org, .net or similar variations and whose content does not require any special configuration to access.

On the other hand, the Deep Web was also part of the web at its conception, and in basic terms, it is the opposite of surface, because its search engines cannot find its content. This is the key difference between the two in real data terms. Sites on the surface Internet are indexed for search engines to find, but the Deep Web is not indexed. However, both are also accessible by the public; they just require different methods to access them – usually a specific password encrypted browser or a set of log-in details. A common image used to represent the meaning of surface versus Deep Web is that of an iceberg: the visible portion of the iceberg represents a very small part of the whole (the whole in this case being the whole of the Internet, surface and Deep Web) as Fig. 1.1 depicts.

The Deep Web contains all of our medical records, financial records, social media files and plenty other important information we want and need to keep secure. It is this need to keep secure files that gave rise to the need to keep a portion of the web secured away from being "googled" at the impulse of anybody at any time.

It is estimated that the Deep Web contains about 102,000 unstructured databases and 348,000 structured databases. In other words, there is a ratio of 3.4 structured data sources for every one (1) unstructured source. Figure 1.1 is the result of a sample of Deep Web databases conducted by Bin et al. (2007) (Fig. 1.2).

Finally, the Dark Web is part of the Deep Web, but it has one major difference. It is not possible to get to the Dark Web using a regular web browser. A special browser is needed, specifically designed for the task such as Tor or similar browser technology. These browsers work differently than conventional browsers, and they are by far the best and most

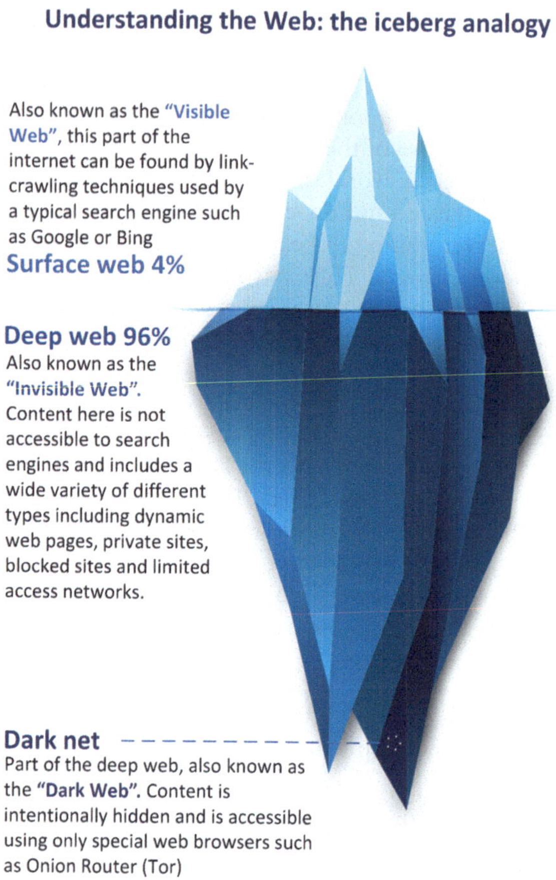

Fig. 1.1 A visual aid in understanding the web (EMCDDA–Europol 2016)

popular (with an estimated 2.5 million daily users). Named "The Onion Router", it was quickly coined with the shorter "Tor" with its name coming from application layer encryption within communication protocol stacks as many layers representing the layers of an onion. With Tor, you'll be able to reach not only the Dark Web but the even smaller subsection known as the Tor network.

	Sampling Results	Total Estimate	99% Confidence Interval
Deep Web sites	126	307.000	236.000 – 377.000
Web databases	190	450.000	366.000 – 535.000
- unstructured	43	102.00	62.000 – 142.000
- structured	147	348.000	275.000 – 423.000
Query interfaces	406	1.258.000	1.097.000 – 1.419.000

Fig. 1.2 Deep Web databases (Bin et al. 2007)

The technology to create the Dark Web was initially created by the US government in the mid-1990s to allow spies and intelligence agencies to anonymously send and receive messages. So easily one can realise that its anonymous nature makes it a good place for all kinds of things people wouldn't dare do on the surface web.

As of 2015, the term "the Dark Web" is often used interchangeably with the Deep Web due to the quantity of hidden services on the darknets. The term is often inaccurately used interchangeably with the Deep Web due to Tor's history as a platform that could not be search-indexed. Mixing uses of both these terms have been described as inaccurate (Bright Planet 2014). The darknet(s) as its name depicts recalls images of shadowy alleys, malicious, hard-faced individuals and socially damaging activities, covering a range from political protestors – rebels – to drug dealers, to terrorist and gun dealers, to paedophiles and everything in between.

Darknets are used for several legitimate purposes: to avoid identity theft, for marketing tracking, to circumvent censorship and to perform research on topics that might be sensitive in certain countries. Chapter 2 will describe both the legitimate and criminal stakeholders of the darknets, as well as their motives behind the use of this side of the web.

Finally, trying to present a definition, the one by Sherman and Price (2007) will be used (as cited in Lievrouw): "[The Dark Web is compromised by] websites that are outdated, broken, abandoned, or inaccessible using standard web browsing techniques". Specifically, the description of "inaccessible" is adequate for our understanding. As we will realise later, many of the sites on the Dark Web strive to be private or at least only accessible to those who know what they're looking for.

Having until now presented synoptically the three web layers, it is possible to present an easy view of the differences among them in Fig. 1.3:

Surface Web	Deep Web	Dark Web
Accessible	Accessible by password, encryption, or through gateway software	Restricted to special browsers
Indexed for Search Engines	Not indexed for Search Engines	Not indexed for Search Engines
Little illegal activity	Little illegal activity outside of Dark Web	Large scale illegal activity
Relatively small in size	Huge in size and growing exponentially	Unmeasurable due to nature

Fig. 1.3 Main web layer differences

1.2 Infrastructure of the Dark Web

As described in the previous section, the Dark Web is intentionally hidden from the general public/simple users. The Dark Web consists of overlay networks, known as darknets, which offer various hidden services. These networks can only be accessed using specific software, such as Tor and I2P (Brown 2016). In this section we will describe the technical infrastructure of the Dark Web through the analysis of the best known darknets and the tools used to access them (Hawkins 2016).

None of the aforementioned software (Tor and I2P) was created to provide safe passage and dissemination of illegal markets and products in neither surface nor Deep Web. Nevertheless, the anonymity and data encryption provided by using software like Tor made them as a tool used by people wanting to sell their illegal services/products (hacking for hire, 0-day vulnerabilities, guns, drugs, etc.), furthermore, for extremist and terrorists who want to disseminate their opinion (propaganda) and proselytise people as well as for people sharing illegal videos and images (e.g. child abuse material (CAM)).

1.2.1 The Tor Project (Tor): Overview

The Onion Routing Project or simply Tor is employing the third generation of the onion routing technique in order to provide anonymous surfing and communication. The onion routing technique (first generation) was developed in the mid-90s by the US Naval Research Laboratory and the Defense Advanced Research Project Agency (DARPA) to provide safe communication between operatives in the field and intelligence gathering (Tor website) by anonymising TCP-based applications. In 2002, the Naval

Laboratory released the source code of Tor, and the Electronic Frontier Foundation (EFF) undertook Tor's founding, becoming the cornerstone for the creation of the Tor Project organisation responsible for maintaining, upgrading and shaping Tor network as it is nowadays (Syverson 2015; Immonen 2016; Çalışkan et al. 2015).

Tor became the most famous tool, in terms of anonymity and privacy to access and publish material on the Dark Web among other tools (e.g. I2P), and thus Tor darknet is the most famous and holds the highest number of visitors and services (Jardine 2015). At the time of writing this chapter and based on The Tor Project (2019b), between 1,500,000 and 3,000,000 relay users (only the ones directly connected to the Tor network) existed, from 3000 up to 7000 relay nodes and more than 1000 bridge nodes, between 60,000/per day and 80,000/per day unique .onion addresses (only version 2) and around 200 Gbits/s bandwidth consumption, while the relays can support approximately 400 Gbits/s overall bandwidth consumption.

As it was described before, services on the Dark Web are intentionally hidden, and thus the .onion sites do not have the formatting used in the clearnet, www.example.com. The .onion top-level domain (TLD) is specifically used to access hidden services hosted only on the Tor network, and it is not part of the Internet DNS root. Furthermore, the addresses under the .onion consist of 16 alphanumeric characters. A user needs to either know the exact address of a hidden service or to use a search engine specifically designed to work on the .onion. A few examples of such sites are below:

1.2.1.1 Search Engines and Introductory Points

TORCH **Torch is a Tor search engine and can be accessed through http://xmh57jrzrnw6insl.onion/**

not Evil **Not Evil is another search engine designed for the Tor network and can be accessed through http://hss3uro2hsxfogfq.onion/**

The Hidden Wiki

The Hidden Wiki is mainly used as a directory of .onion links. The links are categorised based on the service they offer like financial services, drugs, email/messaging and P2P file sharing among others. Nevertheless, one category that it is not included in the site is .onion links regarding terrorism and child sexual abuse material.

Chapter 4 will provide further in-depth details regarding services and markets of the Dark Web as well as the respective links and descriptions, while Chap. 3 will describe the activities of terrorist organisations in the Dark Web and the clearnet and how terrorists are moving from one part of the Internet to the other based on their goals, objectives and activities.

1.2.2 *Tor Architecture and Routing*

Data transmitted using the onion routing is encapsulated in multiple layers of encryption, resembling the layers of an onion (see Fig. 1.2). The number of layers is equal to the number of users acting as nodes, also known as relays, each time. This technique assists the user to remain anonymous and evade eavesdropping and traffic analysis techniques which could reveal the origin, destination and content of a message (The Tor Project 2019c). Tor users have to decide whether they will participate in the network as nodes or not; thus, it is in volunteer bases. As the first generation of onion routing technique, Tor is anonymising TCP packets, while offering significant improvements in comparison with the first generation such as *perfect forward secrecy, separation of "protocol cleaning" from anonymity, many TCP streams can share one circuit, leaky-pipe circuit topology, congestion control, directory servers, variable exit policies, end-to-end integrity checking, rendezvous points and hidden cervices* (Dingledine et al. 2004) (Fig. 1.4).

Tor consists mainly of (i) the Tor browser which offers the appropriate setting (e.g. proxy) in order to connect to the Tor network and (ii) the hidden services/sites hosted in the Tor network. Furthermore, Tor network consists of the Tor nodes and the directory servers, while the nodes share part of their bandwidth meaning that it will increase or decrease based on the number of the nodes; thus the higher the number of nodes in Tor, the faster it will be (The Tor Project 2019e).

1.2.2.1 *Tor Nodes*

Tor nodes, or relays, are created by users offering their computer(s), purely on a volunteer basis, in order to be used as a node. It is highly important

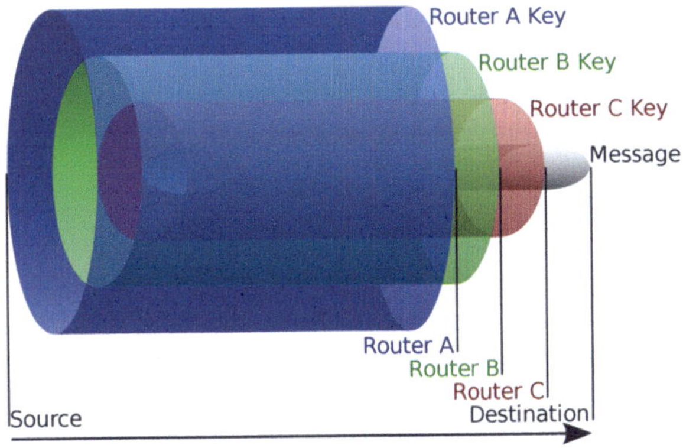

Fig. 1.4 Onion routing technique (Neal 2008)

to explain that the higher number of nodes means higher available bandwidth, increased robustness of the network against attacks and making it more difficult to analyse the traffic. Tor consists of three types of nodes based on (Erkkonen et al. 2007; Aschmann et al. 2017; The Tor Project a, b, c, d, e; Tor Challenge 2018):

- The guard/entry node: the guard node is the first node each user will hop to in order to connect to the Tor network and to the requested service/site. The selection of the guard nodes is done at the user level, and the selection is random in order to minimise the chances of "eavesdropping".
- Middle or internal node: middle nodes are the nodes that exist between the guard node and the exit node.
- Exit nodes: exit nodes are the last nodes before a user reaches the requested destination, and thus this type of node is responsible for sending the request either out of the Tor network or to a hidden service.
- Bridge node: the main difference with the aforementioned nodes is that bridges are not listed in the main Tor directory authority. Thus, it will be difficult for ISPs to block Tor traffic passing through these bridges. At the moment of writing this section, Tor has more than 1500 IPv4 bridges and around 250 IPv6 bridges (The Tor Project 2019d).

To run nodes might have legal impact, especially for the ones running exit node, as it can be used to identify the respective IP address if the exit node is used for illegal purposes. Furthermore, it is advisable not to run exit node using a home computer while it is best to notify the ISP (The Tor Project 2018a).

1.2.2.2 Tor Directory Authorities

The Tor directory authorities, which are ten (see Fig. 1.5) at the time of writing this chapter, are databases which contain information and the list of all the active nodes at the network. Thus, they have complete knowledge and view of the network's topology. Information relevant to the routers stored in the director authorities is encrypted with digital signatures. To further secure both the directory authorities as well as the entire Tor network, the administrator of each server will process and approve information regarding nodes in order to be published to users (Erkkonen et al. 2007).

1.2.2.3 Tor Circuit

In order for a user to connect to the Tor network, the user's Tor client (Tor browser) will have to communicate with a Tor directory authority, holding a list of all the available Tor nodes. Once the user receives this list,

Nickname†	Advertised Bandwidth	Uptime	Country	IPv4	IPv6
● dizum (2)	3.13 MiB/s	28d 1h		45.66.33.45	-
● Serge (1)	1.45 MiB/s	6d 2h		66.111.2.131	2610:1c0:0:5::131
● moria1 (1)	500 KiB/s	11d 20h		128.31.0.34	-
● tor26 (1)	75 KiB/s	9d 6h		86.59.21.38	2001:858:2:2:aabb:0:563b:1526
● bastet (1)	50 KiB/s	2d 3h		204.13.164.118	2620:13:4000:6000::1000:118
● maatuska (8)	50 KiB/s	20d 15h		171.25.193.9	2001:67c:289c::9
● dannenberg (1)	40 KiB/s	22d 6h		193.23.244.244	2001:678:558:1000::244
● Faravahar (1)	40 KiB/s	8d 22h		154.35.175.225	2607:8500:154::3
● gabelmoo (1)	40 KiB/s	26d 7h		131.188.40.189	2001:638:a000:4140::ffff:189
● longclaw (1)	38 KiB/s	9d 21m		199.58.81.140	-
Total	**5.4 MiB/s**				

Fig. 1.5 Tor directory authorities (29/11/209) (Tor Metrics 2019)

the client will randomly decide the path of nodes that will be used in order to reach the destination server, which in turn publishes the hidden service the person is looking for (see Fig. 1.6) (Aschmann et al. 2017; The Tor Project 2018a, b). Each node has knowledge only for the previous and next node (one hop knowledge) of the network, exchanging a different encryption key each time. This method ensures that even if one node is compromised it will not be able to identify the entire path of the Tor network. In order for a route, or circuit, to be formed, the Tor browser will download the current list of register nodes, from the directory authorities, and it will randomly select a guard node. Then, it will select the rest of the nodes based on their bandwidth and stability (highest bandwidth and only stable nodes are selected). By default, when three nodes are selected, the circuit is formed, and the generation of encryption keys follows.

1.2.2.4 Setting Up Tor

Tor is available for Windows, Apple MacOS and GNU/Linux, and it can be downloaded from https://www.torproject.org/projects/torbrowser.html.en in sixteen (16) languages:

- English
- Arabic
- Deutsch

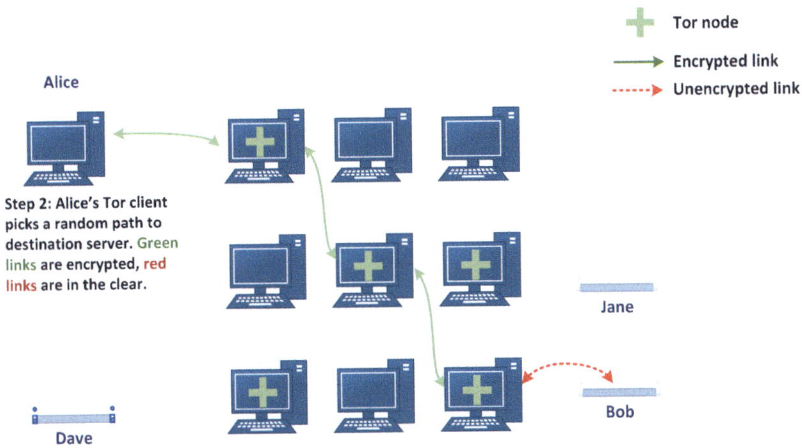

Fig. 1.6 Tor network-circuit setup (The Tor Project 2019a)

- Spanish
- Farsi
- French
- Italian
- Japanese
- Korean
- Dutch
- Polish
- Portuguese
- Russian
- Turkish
- Vietnamese
- Chinese

To install Tor in Windows machines is a straight forward procedure. The user has to download the appropriate file from the list, save the file and then open it. Choose "Run", choose your language of preference, and press install. When the installation is complete, press finish, run the Tor browser, and click "Connect". To install Tor in MacOS, the user needs to download the respective file, save it and drag the *.dmg* file into the application folder.

To install Tor in Linux/GNU, the user needs to first download the architecture file and then run the following commands from the terminal: tar-xvJf tor-browser-linux32-7.5.3_*LANG*.tar.xz (for 32-bit OS) or tar-xvJf tor-browser-linux64-7.5.3_*LANG*.tar.xz (for 32-bit OS). The next step is to navigate to the Tor browser using the following command: cd tor-browser_*LANG*, and run the Tor browser either from the graphical interface (click it) or by executing the following command: ./start-tor-browser.desktop from the terminal.

Finally, Tor browser can also be installed in android-based machines such as smartphones and tablets from Google Play.

1.2.3 The Invisible Internet Project (I2P): Overview

In this section we are describing the I2P network and its infrastructure. I2P is a decentralised, peer-to-peer overlay network that started in 2003, offering anonymity by employing the *garlic routing/encryption* techniques (Erkkonen et al. 2007). The design of the network is message based in order to run on top of IP, but communication can also be achieved on top of TCP and UDP based on the requirements of each application/service.

The I2P client software can act as a router once it is installed in a machine, providing connectivity to I2P websites (TLD, .i2p) in the darknet, or it can host a service (e.g. an .i2p website).

The garlic routing technique, which is a variant of onion routing, was coined back in 2000 and in the framework of I2P provides the following three attributes (I2P Garlic Routing 2018):

- Tunnel building and routing (in order to transmit data, each router creates one-way tunnels (inbound and outbound tunnels)).
- Data bundling to be able to evaluate the end-to-end message delivery status.
- ElGamal/EAS + SessionTags encryption algorithms are used to provide end-to-end encryption and minimise the possibility of traffic analysis attacks.

I2P can be downloaded from geti2p.net/en/download, and it is available for Windows, Mac OSX, GNU/Linux, BSD, Solaris, Debian, Ubuntu and Android.

1.2.4 I2P Network Database

In order for a client to connect to other clients and setup a circuit, it will have to ask the I2P netDB which contains all mandatory information regarding other user's inbound tunnels. The I2P netDB basically contains two important types of records, the RouterInfos, which is the contact information of the I2P routers (IP address, respective port and public key), and the LeaseSets, which contains the destination contact information (tunnel endpoints and the public key of the requested service). Furthermore, tunnels expire every 10 minutes; thus clients have to request the aforementioned information from the netDB, if they want to stay connected with the service (Egger et al. 2013; I2P, 2018b).

1.2.5 I2P Routers and Tunnels

I2P routers use two pairs of one-way tunnels in total from which the one handles the inbound traffic and the other the outbound traffic as Fig. 1.7 depicts. Thus, for one message and the respective reply, the router will build four tunnels each time. To clarify how the I2P tunnels work, we first need to understand the philosophy of how the inbound and outbound tunnels are built. The creator of a tunnel decides the number of the peers

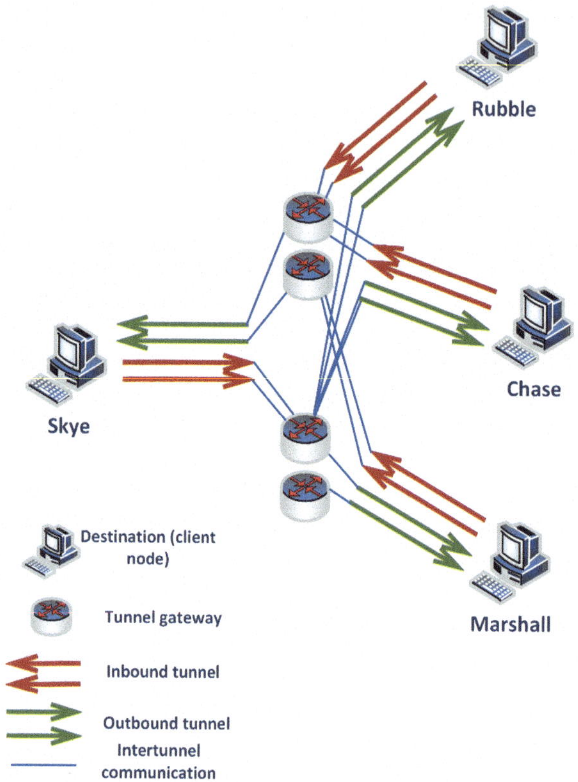

Fig. 1.7 I2P tunnels

(number of hops) and which peers will participate in the tunnel in order to strengthen the security of the tunnel and minimise any chances of either third parties or other tunnel participants to identify the total number of hops the tunnel has and if they belong in the same tunnel (Erkkonen et al. 2007; Egger et al. 2013; I2P 2018a).

1.2.6 Freenet: Overview

Freenet developed in 1999 and released in 2000 is most likely the third ranking darknet after Tor and I2P in terms of the number of users (Brown 2016). It is a peer-to-peer network designed for sharing, storing and retrieving files as well as publishing Freenet websites, called "freesites" providing high anonymity to the users. Furthermore, it is not a centralised platform; thus it has stronger resistance in attacks. From the release of 0.7.5 version, the architecture of the Freenet has changed, strengthening user's privacy and enhancing the security of the Freenet nodes against malicious attacks – see Freenet's architecture and design – (Clarke et al. 2001, 2010). As it is a decentralised network, Freenet relies on its users to store, insert, edit and request files anonymously. To achieve that, it is "mandatory" for all Freenet users to contribute in terms of hard drive space (portion of their own hard drive) and their bandwidth. These files can be anonymous or pseudonymous static websites, forums, microblogs and regular files. The five main goals of the design followed by Freenet designers based on Clarke et al. (2001) are:

1. Anonymity for both producers and consumers of information
2. Deniability for storers of information
3. Resistance to attempts by third parties to deny access to information
4. Efficient dynamic storage and routing of information
5. Decentralisation of all network functions

Freenet is free for everyone and can be downloaded from freenetproject.org/pages/download.html and is available for Windows, GNU/Linux, Mac OSX and Posix.

1.2.6.1 Freenet's Architecture and Design

As discussed in the previous section, Freenet is a peer-to-peer overlay network consisted from as many nodes as its users. Each node provides a part of their hard drive and bandwidth for storing, retrieving and editing Freenet files. These files are stored after they are divided into encrypted blocks, distributed among multiple nodes, while the holders of the files are not familiar with the content of the files (Aschmann et al. 2017). The stored files are associated with a key (or address) which is based on a string, given by the user. This key has two purposes, to locate where in the Freenet network the file is stored and to authenticate this file.

Up until the release of 0.7.5 version, Freenet was designed to choose the edges and the nodes of the network to be used based on the best optimisation route. Version 0.7.5 introduced the darknet mode which allows the nodes and edges to connect only to nodes the user trusts (friends list), with whom they have previously exchanged public keys as Fig. 1.8 depicts. Thus, the new architecture offers two choices to the users, either to use

Opennet mode	Darknet Mode
Easy to block	Very hard to block
Limited anonymity	Good anonymity
Somewhat centralised	Fully decentralised

Fig. 1.8 Darknet vs opennet mode

the darknet mode and stay hidden by connecting only to trusted nodes (creation of private network) or to also connect to nodes operated by strangers (opennet mode) (Aschmann et al. 2017; Clarke et al. 2010). Figure 1.8 summarises the key differences between Freenet's darknet and opennet mode (Freenet Project, 2018b):

Freenet nodes are designed to cache as many of the files they transfer as possible; thus the node's storage is getting fuller easier and faster. To this extent, Freenet is using a best-effort algorithm in order to randomly remove files from a node when its storage is full (Fig. 1.9).

1.2.6.2 Sharing, Requesting and Accessing Data

As it was described in the previous section, each file (or part of it) is associated with a key. All nodes have a routing table depicting the addresses and their respective keys. Thus, if a user is "looking" for a file, he has to send a request based on the key of the file. To locate the file, the request hops from node to node (creating a route) based on which neighbour node is the most desirable, from those available. When this is not feasible, the request goes back to the previous node in order to restart the route as depicted in Fig. 1.10. When the file is located within a node, then the file is backtracked to the previous node which forwards it back to the node that initially requested the file. This process is terminated either because the file has been found or because the nodes gave up looking for the file (Clarke et al. 2010).

To share and store a file, a user will have to "upload" the file, and then it will be cut down to smaller parts in order to be distributed among the nodes, following the same logic with the one followed when a user is requesting a file. First, the file will be associated with a key, and then it will create a route, and when it is terminated, the file will be forwarded to the node that terminated the route. The user that uploaded the file does not have to stay online as the file has been stored in other nodes on the network (Balduzzi and Ciancaglini 2015).

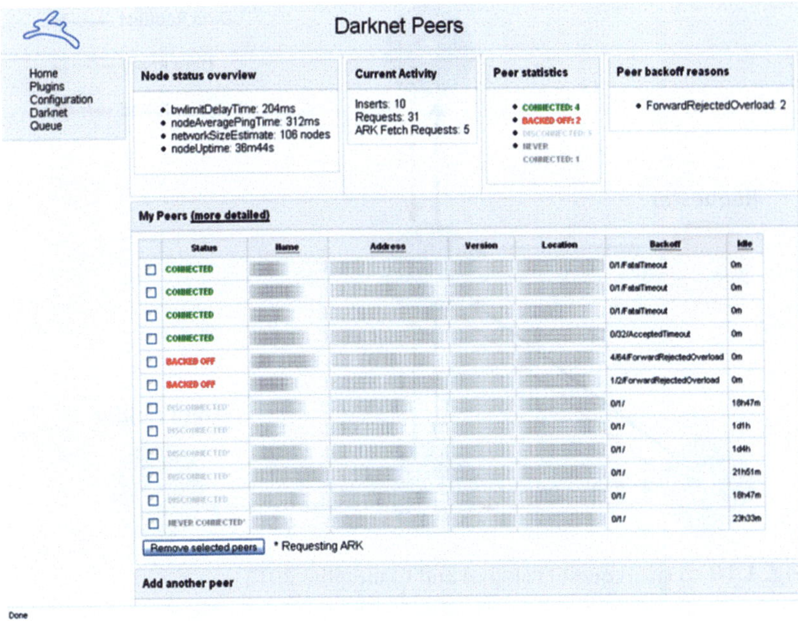

Fig. 1.9 Freenet darknet mode peers (Freenet Project 2018a)

A user can search and access documents or freesites by either using a Freenet search engine such as Freegle or by typing the key associated with the data/freesites in the following format, http://localhost:888/[Freenet Key].

For example, if you have installed Freenet to your pc, you can insert the following key, http://127.0.0.1:8888/freenet:USK@tiYrPDh~fDeH5V7NZjpp~QuubaHwgks88iwlRXXLLWA,yboLMwX1dChz8fWKjmbdtl38HR5uiCOdIUT86ohUyRg,AQACAAE/nerdageddon/-49/ in order to download the Nerdageddon freesite which includes most of the freesites on Freenet, excluding most of the pornographic freesites. Furthermore, freesites do not contain dynamic content (e.g. scripts and databases), and they are constructed in HTML.

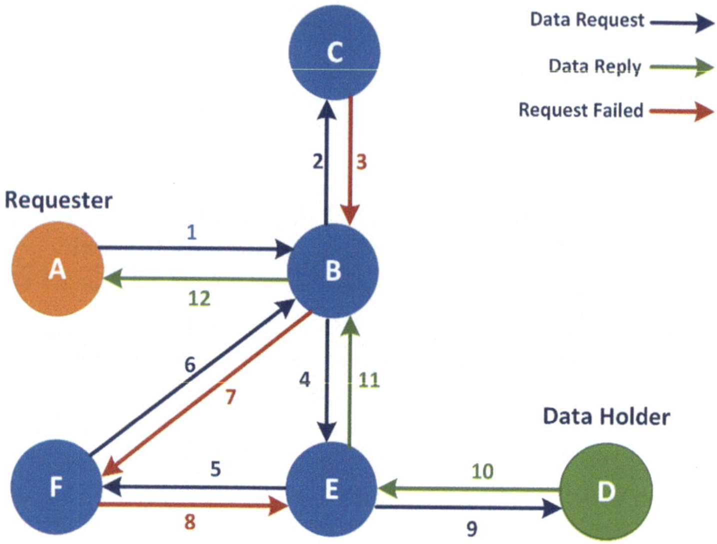

Fig. 1.10 Data request (Balduzzi and Ciancaglini 2015)

1.3 Law Enforcement Agency Investigating the Dark Web

The above-mentioned definitions guide us to the area where the exact problems are encountered, especially in the Deep and Dark Web during a law enforcement agency investigation. The use LEA's make of the Dark Web itself for investigations is usually not specified in the openly available literature. It is nevertheless generally known that the activities on the Dark Web are an object of investigation, within two operational contexts, as listed below:

- Proactive investigation for intelligence, employing the use of Tor and often manual searches of the encountered content; in a few cases, LEAs are experimenting the use of novel automatic tools to crawl portions of the Dark Web and index their content.
- Reactive investigations, for example, to collect information on the Dark Web on a suspect or organisation. These operations are also challenging, mainly because of the difficulty to de-anonymise digital traces on the Dark Web.

The basic conclusion that derives from all the above is that an anonymity-granting system such as Tor, along with other similar technologies, is just a tool. The endurance and impact of operations concerning the Dark Web are often mentioned in the media and in LEA's official reports, so in the following paragraphs, we will indicatively focus on "how policing the clear web influences the Dark Web" and on the "policing dilemma the Dark Web poses".

1.3.1 How Policing the Clear Web Influences the Dark Web

Online policing is also as beneficial as offline policing. The anonymity of Tor does not necessarily slow down law enforcement efforts. There are limits to the effectiveness of online policing. One limitation is that online criminals can be global, even while most law enforcement agencies (Interpol not included) are local. Another limitation is that cybercrime is rapidly increasing, which threatens to overwhelm any and all available policing capacity of nations.

Policy actions might assist LEAs to overcome the aforementioned limitations, but before considering them, it is important to highlight that policing actions on the surface web influence the Dark Web. An example is given by the Internet content regulation from a drug-policy perspective: measures such as the Australian compulsory Internet filtering regime to block drug contents on clear web websites would likely drive drug discussions to the Deep or the Dark Web, where digital spaces are not affected by Internet filtering and where governments are unable to regulate Tor website content (Barratt et al. 2013). On the other side, this measure might also push violent online extremism into the Dark Web, where monitoring of content is much more difficult and less debate takes place (Hussain and Saltman 2014).

1.3.2 The Dark Web Also Poses a "Policing Dilemma"

Based on Jardine (2015), anonymity can be the shield for people doing "good" and for people that without it surfing the web could be impossible. As it was previously discussed, Tor, I2P, Freenet, etc. are just technologies, tools designed neither of ill nor good use. Thus, it is the person and the reason behind them who are responsible for the deeds.

Based on the available literature, there are few examples regarding the difference between Tor technology and the Tor hidden services, with the latter based on Guitton (2013) promoting unethical and illegal content and strongly believing that a stop should be put on the development of Tor hidden services. Other sources (Jardine 2015) argue that shutting down anonymity networks will damage greatly those people that need this technology and use it within legal boundaries. Furthermore, based on Stevens and Jardine (Stevens 2009; Jardine 2015), actions and methods should be employed to raise awareness regarding technologies granting anonymity and how people around the world benefit from them. After all what matters is not what the technology is, but how it is used and what the net effect turns out to be. Based on Brink et al. (2016):

> When it comes to the Dark Web similar attitudes appear across law enforcements worldwide: authorities tend to focus on attacking the offender. The effectiveness of this approach is questioned in the academic literature. Two points are also highlighted: the difference between various cybercrimes perpetuated through the Dark Web might call for different type of measures; moreover, measures developed to exercise control on phenomena on the Clear Web might bring weak if not counterproductive effects if applied to the Dark Web, as already mentioned.

LEAs face a great challenge in the fight against criminal activities on Tor hidden services (THS). As previously explained, Tor is designed so that no single entity in the circuit, including the ISPs, has knowledge of the complete circuit, meaning that it is highly difficult to monitor the website a user is visiting and the user's behaviour in general (Griffiths 2013). However, Dark Web anonymity can be used by LEAs in order to maintain anonymous communication with their sources and undercover agents and to conduct online sting operations and surveillances. Furthermore, LEAs were able to achieve significant wins against cybercriminals in the Dark Web (Nath and Kriechbaumer 2015; Brink et al. 2016):

- Malicious software attacks in August 2013, computers of people accessing THS hosted by servers of the company Freedom Host were infected by malicious software.
- Silk Road shutdown in October 2013, the FBI arrested the alleged operator of the illegal marketplace Silk Road, which was operated as a THS.

- Operation Onymous about a month after Silk Road was taken offline, the THS Silk Road 2.0 was launched to continue the criminal market activities.

There hasn't been any publicly available (formal) reports discussing and explaining the strategy, methods and tools used to achieve the aforementioned shake downs and apprehension of the cybercriminals. Two methods that may have been employed are:

- Exploitation of user mistakes
- Exploitation of Tor technical limitations and security issues

It is also worth mentioning the operations of other actors, who autonomously infiltrate, disrupt and eventually take down Dark Web websites and services (Weimann 2015; Brink et al. 2016):

- The self-named "Operation Darknet" by Anonymous. In October 2011 they released in the public around 1500 user credentials from a CAM website in the Dark Web.
- The self-named "Operation Paris", by Anonymous, took down hundreds of ISIS websites on the surface web. ISIS had to move to the Dark Web after this operation, and al-Hayat Media Center posted a link and guidelines on how to reach their new site in the Dark Web.

1.4 Concluding Remarks

The exponential growth of sites offering illegal goods (e.g. guns, drugs, etc.), services (e.g. hacking as a service), closed forums and sites relevant to radicalisation, extremism and terrorism combined with the anonymity the darknets provide poses continuing challenges to the LEAs, since Dark Web sites proliferate at a rate far greater than LEAs have been able to intervene. Also, current tools seem to not be fully adequate to counter activities taking place in the Dark Web, while cybercriminal's quiver regarding tools, techniques and methodologies is evolving continuously. The right of anonymity every person has alongside with the jurisdiction challenges posed by the nature of the Internet multiplies the level of difficulty for those responsible to safeguard society. Finally, it might be difficult to justify the cost and the person months of operations aimed at regulating and monitoring the Dark Web, while at the same time, LEAs have to also fight against other forms of cybercrime.

In response, LEAs are constantly looking for technological tools to assist the identification and analysis of illegal material on the Dark Web, as well as the proactive identification of radicals, extremists and terrorists, while at the same time respect the anonymity of legitimate users. The TENSOR platform designed to meet the aforementioned objectives consists of a number of tools that enable the proactive identification, collection and analysis of terrorist-related content.

References

M. Aschmann, L. Leenen, J. Jansen van Vuuren, *The Utilisation of the Deep Web for Military Counter Terrorist Operations.* (Academic Conferences and publishing limited, s.l., 2017)

M. Balduzzi, V. Ciancaglini, *Cybercrime in the Deep Web* (Black Hat Europe, Amsterdam, 2015)

M.J. Barratt, S. Lenton, M. Allen, Internet content regulation, public drug websites and the growth in hidden internet services. Drugs **20**(3), 195–202 (2013)

H. Bin, M. Patel, Z. Zhang, Accessing the deep web: A suervey. Commun. ACM **50**, 94–101 (2007)

Bright Planet, *Clearing Up Confusion – Deep Web vs. Dark Web.* [Online] (2014), Available at: https://brightplanet.com/2014/03/clearing-confusion-deep-web-vs-dark-web/

P. Brink et al., *MEDIA4SEC Report on State of the Art Review.* (MEDIA4SEC, s.l., 2016)

B. Brown, *Threat Advisory: 2016 State of the Dark Web.* (AKAMAI Threat Advisory, s.l., 2016)

E. Çalışkan, T. Minárik, A.-M. Osula, *Technical and Legal Overview of the Tor Anonymity Network.* (NATO Cooperative Cyber Defence Centre of Excellence, s.l., 2015)

I. Clarke, O. Sandberg, B. Wiley, T.W. Hong, Freenet: A distributed anonymous information storage and retrieval system, in *Designing Privacy Enhancing Technologies*, ed. by H. Federrath, (Springer, Berlin, Heidelberg, 2001), pp. 46–66

I. Clarke, O. Sandberg, M. Toseland, V. Verendel, *Private Communication Through a Network of Trusted Connections: The Dark Freenet.* (Network, s.l., 2010)

R. Dingledine, N. Mathewson, S. Paul, *Tor: The Second-Generation Onion Router* (Naval Research Lab, Washington, DC, 2004)

C. Egger, J. Schlumberger, C. Kruegel, G. Vigna, in Practical attacks against the I2P network, ed. by A. S. C. V. W, S. J. Stolfo. *Research in Attacks, Intrusions, and Defences* (Springer-Verlag Berlin Heidelberg, Berlin, 2013), pp. 432–451

EMCDDA–Europol, *EU Drug Markets Report: In-Depth Analysis (European Monitoring Centre for Drugs and Drug Addiction and Europol)* (Publications Office of the European Union, Luxembourg, 2016)

H. Erkkonen, J. Larsson, C. Datateknik, *Anonymous Networks.* (Computer communication and distributed systems, s.l., 2007)

Freenet project, *Documentation.* [Online] (2018a), Available at: https://freenetproject.org/pages/documentation.html. Accessed on 26 Mar 2018

Freenet project, *Freenet Help.* [Online] (2018b), Available at: https://freenetproject.org/pages/help.html. Accessed on 26 Feb 2018

M. Griffiths, *Monitoring Internet Communications* (POST – Parliamentary Office of Science and Technology, London, 2013)

C. Guitton, A review of the available content on Tor hidden services: The case against further development. Comput. Hum. Behav. **29**(6), 2805–2815 (2013)

B. Hawkins, *Under The Ocean of the Internet – The Deep Web* (SANS Institute-InfoSec Reading Room, 2016)

G. Hussain, E.M. Saltman, *Jihad Trending: A Comprehensive Analysis of Online Extremism and How to Counter It* (Quilliam, s.l., 2014)

I2P, *i2p Tunnel Implementation.* [Online] (2018a), Available at: https://geti2p.net/en/docs/tunnels/implementation. Accessed on 23 Mar 2018

I2P, *The Network Database.* [Online] (2018b), Available at: https://geti2p.net/en/docs/how/network-database. Accessed on 23 Mar 2018

I2P Garlic Routing, *Garlic Routing and "Garlic" Terminology.* [Online] (2018), Available at: https://geti2p.net/en/docs/how/garlic-routing. Accessed on 22 Feb 2018

V.V. Immonen, *Alice in Onion Land: On Information Security of Tor* (ITA-SUOMEN YLIOPISTO, s.l., 2016)

E. Jardine, The Dark Web Dilemma: Tor, Anonymity and Online Policing. *Global Commission on Internet Governance Paper Series,* Band 21 (2015)

C. Nath, T. Kriechbaumer, *The Darknet and Online Anonymity* (POST – Parliamentary Office of Science and Technology, London, 2015)

H. Neal, *Wikimedia commons.* [Online] (2008), Available at: https://commons.wikimedia.org/wiki/File:Onion_diagram.svg. Accessed on Apr 2018

C. Sherman, G. Price, *The Invisible Web: Uncovering Information Sources Search Engines Can't See* (Information Today, Medford, 2007)

T. Stevens, Regulating the 'Dark web': How a two-fold approach can tackle peer-to-peer radicalisation. RUSI J. **154**(2), 28–33 (2009)

J. Strickland, *Who owns the Internet?* [Online] (2014), Available at: https://computer.howstuffworks.com/internet/basics/who-owns-internet.htm

Syverson P, *Basic Course on Onion Routing* (U.S. Naval Research Laboratory, s.l., 2015)

The Tor Project, *The Legal FAQ for Tor Relay Operators.* [Online] (2018a), Available at: https://www.torproject.org/eff/tor-legal-faq.html.en. Accessed on 20 Nov 2019

The Tor Project, *Tor Project: FAQ.* [Online] (2018b), Available at: https://www.torproject.org/docs/faq.html.en#EntryGuards. Accessed on 20 Nov 2019

The Tor Project, *The Solution: A Distributed, Anonymous Network.* [Online] (2019a), Available at: https://www.torproject.org/about/overview.html.en#thesolution. Accessed on 21 Mar 2019

The Tor Project, *Tor Metrics.* [Online] (2019b), Available at: https://metrics.torproject.org/. Accessed on 20 Nov 2019

The Tor Project, *Tor Project.* [Online] (2019c), Available at: https://www.torproject.org/. Accessed on 20 Nov 2019

The Tor Project, *Tor: Bridges.* [Online] (2019d), Available at: https://www.torproject.org/docs/bridges.html.en. Accessed on 21 Mar 2019

The Tor Project, *Tor Relay Guide – Tor Bug Tracker & Wiki.* [Online] (2019e), Available at: https://trac.torproject.org/projects/tor/wiki/TorRelayGuide. Accessed on 22 Feb 2019

Tor Challenge, *What Is a Tor Relay?* [Online] (2018), Available at: https://www.eff.org/torchallenge/what-is-tor.html. Accessed on 21 Mar 2018

Tor Metrics, *Relay Search-Flag: Authority.* [Online] (2019), Available at: https://metrics.torproject.org/rs.html#search/flag:authority. Accessed on 29 Nov 2019

G. Weimann, Going dark: Terrorism on the dark web. Stud. Confl. Terror. **39**(3), 195–206 (2015)

CHAPTER 2

Using the Dark Web

Dimitrios Kavallieros, Dimitrios Myttas, Emmanouil Kermitsis, Euthimios Lissaris, Georgios Giataganas, and Eleni Darra

2.1 Introduction

The Dark Web and the underlying darknets (e.g. Tor) provide high levels of anonymity and security due to their architecture and the reason the Dark Web initially designed for. As it was previously mentioned, the Dark Web was developed to provide anonymity and safety to spies and field agents. Nevertheless, this philosophy and the actual architecture of the darknets provided a safe environment for criminal activities.

D. Kavallieros (✉)
Center for Security Studies-KEMEA, Athens, Greece

University of Peloponnese-Department of Informatics and Telecommunications, Tripoli, Greece
e-mail: d.kavallieros@kemea-research.gr

D. Myttas • E. Kermitsis • E. Lissaris • G. Giataganas • E. Darra
Center for Security Studies-KEMEA, Athens, Greece
e-mail: d.myttas@kemea-research.gr; e.kermitsis@kemea-research.gr; e.lissaris@kemea-research.gr; g.giataganas@kemea-research.gr; e.darra@kemea-research.gr

B. Akhgar et al. (eds.), *Dark Web Investigation*,
Security Informatics and Law Enforcement,
https://doi.org/10.1007/978-3-030-55343-2_2

It is highly difficult to accurately classify the sites in the different darknets due to the structure. Most studies have been focused on Tor as it is the most well know and used darknet. According to a report published by Intelliagg (2016), there were more legal websites than illegal in Tor under the UK and USA. These findings were based on machine learning algorithms which were responsible for the classification of the data, as Fig. 2.1 depicts, while based on manual classification (1000 websites) the research found that 68% of the sites contained illegal material. To reach to these conclusions, 13,584 websites in total were analysed.

Due to the inherent properties of Tor, a quantitative analysis of the use of Tor to establish a ratio between the legal and illicit use is very challenging. There is no academic agreement on this question, and no hard numbers are presented, while based on Biryukov et al. (2014) the ratio between legal and illegal content is 50:50. As it is highly difficult to accurately classify the content of the Dark Web, it is certain that two generic categories of actors visit it, the non-criminal and criminal.

The double nature of the Dark Web, where legitimate activities take place in parallel to criminal ones, requires a balance between individual freedoms, such as freedom of speech, and the need to fight crime. We can therefore see the Dark Web used for several legitimate purposes as an example of the 'good', while for its criminal purposes both as the 'bad' and the 'ugly' examples of the digital world of social media and although based on digital communication and digital transactions, it may have a strong negative effect on the physical public safety and security (MEDIA4SEC-TNO, Serena Oggero 2017).

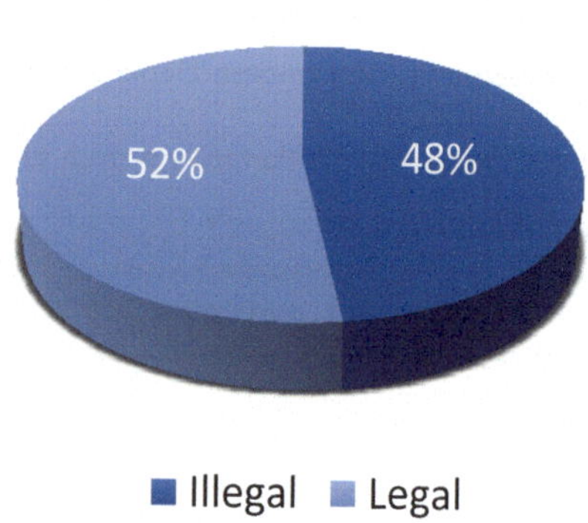

Fig. 2.1 Tor website – legal vs illegal (Intelliagg 2016)

2.2 Dark Web Users

As it was thoroughly described in Chap. 1, darknets (and the software used to grant access) provide great anonymity and communication. None of the darknets are created to facilitate illicit actions but to provide secure communication between legit users (e.g. field operatives, journalists and whistle-blowers). Nevertheless, technology itself is agnostic in terms of usage and depends on the type of person using it (and for which purpose). Criminal actors took advantage of the 'safe' passage this technology offers, and therefore, users of the Dark Web are classified into two major categories:

- The non-criminal users utilising the anonymity and security advantages of Dark Web
- The criminal actors carrying out illegal activities facilitated by the Dark Web environment

Legitimate users move from the surface to the Dark Web to protect their identities, sources and privacy through the anonymity, security and safety provided in the Dark Web environment, whereas at the same time criminals take the same advantages as weapons to act illegally in the same space.

Typical examples of legitimate users using the Dark Web, in addition to the public of ordinary users, are listed in Fig. 2.2.

Dark Web User Groups	Purpose of Beneficial Use
(Political) activists and whistle-blowers	Operate anonymously in totalitarian regimes; expose business or government related injustices reducing the risk of repercussions
Journalists	Protect sources and themselves while publishing non-state-controlled articles
Law enforcement agents (LEAs)	Receive truly anonymous tips, use the internet during surveillance, protect undercover staff
Businesses	Support corporate spying and market screening operations
The military	Share confidential data, protect the identity of field agents, gather intelligence

Fig. 2.2 Non-criminal user groups

2.2.1 *Non-criminal Actors*

There are many groups of legitimate users from different sectors, including those policing the criminality, who become users of the Dark Web for specific purposes. By taking advantage of its features, there are seven typical groups of legitimate users identified and listed in the following sections.

2.2.1.1 Public-Ordinary Users

Normal people use the Dark Web for various reasons but mainly to browse the web without being tracked, thus protecting their privacy to be exposed in various risks. The Tor Project lists some of these reasons below and recommends users to make use of Tor browser (or other browsers such as those described in Chap. 1) for these cases (TOR 2018):

- To protect their privacy from personal data
- Secure and private communication with organisations which are giving away personal information and mailing lists to third parties
- To protect their current geolocation through revealing their IP
- To have access to sensitive research topics (e.g. religion) for which in some countries is restricted in the surface web
- To have access to the Internet and social media platforms such as Facebook or YouTube that from time to time are being blocked by the governments in some countries
- To protect their privacy from online surveillance and monitoring by making it extremely difficult for an observer to correlate the visiting sites with the physical-world identity

2.2.1.2 Journalists

Journalists nowadays are using the Dark Web to avoid state censorship and on occasions arrest. At the same time, their sources are better protected due to the anonymity darknets offer. Furthermore, journalism faces other key issues such as (Murray 2014; CIVIL 2018; Con 2018):

- The oligarchy of stakeholders involved in the shaping of common opinion via the traditional media
- The freedom of speech in sensitive topics putting the privacy of journalists and their sources at risk
- The more and more expanding number of uncertified journalists via the new digital media

- The truth, trust and verification of these increasing journalist's outputs in the new media
- The financial sustainability of the journalism itself in the new digital landscape and economy

Freedom of speech, privacy and truth in journalism, being under attack nowadays through mainstream media, make journalists and their audience use the Dark Web for the following main reasons (RFS 2018):

- To protect journalists' identity and sources and ensure their privacy and safety, while publishing non-state-controlled articles. Journalists, sources, bloggers and dissidents from all over the world, being Internet prisoners of conscience, jailed or harmed, are advised, for example, by Reporters Without Borders to use Tor.
- To enable both journalists and their audience to overcome the 'national firewalls' and the surveillance of repressive regimes imposed in some countries giving them the chance to write and read, respectively, controversial topics related to democracy, economics, religion, etc. and promoting, opposing and encouraging viewpoints on social changes and political reforms

There is also another role for journalists and media professionals in general, and that is to educate and sensitise the larger public on crime on the Dark Web and potentially to undermine its culture of trust (MEDIA4SEC-TNO, Serena Oggero 2017).

2.2.1.3 *Activists*

Apart from the journalism and the media actors, many of which could also be considered as part of the activist groups, those using darknets are:

- Human rights organisations
- Social movements
- Political activists

Human rights activists are provided with the ability to raise their voice and report anonymously work-labour and other types of abuses to organise people in accordance with the Universal Declaration of Human Rights. The Global Voices and the Human Rights Watch highly recommend the use of browsers providing great anonymity such as Tor for the aforementioned purposes in order to avoid censorship.

2.2.1.4 *Businesses and IT Industry*

Corporate use of the Dark Web is made for two major reasons: either to make dark business or to protect their corporations, resources, customers and products.

The first part is covered in more detail in one of the following chapters of this book dedicated to darknet markets and in the section of this chapter for the criminal actors of the Dark Web. For the second part, the relevant IT departments (e.g. security specialists, SOC teams, security incident response, etc.) have to familiarise themselves with the Dark Web in order to successfully monitor the darknet and its markets regarding new computer vulnerabilities, exploits and hacking tools. This information will assist them to take the appropriate countermeasures limiting their exposure to the associated risks.

Large corporations face today many cyberthreats coming from their internal or external environment. Some of these corporate threats include but not limited to phishing attacks, malware, distributed denial-of-service attacks (DDoS) and hacking attacks leading to corporate sensitive data and digital assets stolen and placed for sale or extortion purposes in the darknet.

Apart from the corporate data and identity information, corporate assets such as intellectual properties, software and digital products are stolen and appearing more and more in the darknet marketplaces for sale and ransomware.

In 2014, hackers stole the data of 500 million users from Yahoo, including names, email addresses, telephone numbers, birth dates and encrypted passwords, with many users having also built around the free email services their digital identities from their bank accounts to photo albums. A similar case in 2016 was the uTorrent breach (data stolen of 400,000 accounts). In both cases stolen data appeared on the darknet's illicit marketplaces (Delamarter 2016). The point of stealing all that data, of course, was to make money from it, and the darknet tells you how to do that (Glick 2016).

Another great example to describe the aforementioned is the healthcare IT industry, one of the most globally growing industries. In 2016, InfoArmor, an underground research team, detected proprietary source code of PilotFish Technology, a healthcare software vendor, which appeared for sale on the popular darknet marketplace AlphaBay. In addition to source codes, some compilation instructions have the specific usernames of PilotFish employees, creating further risk to the organisation. The threat actor claimed also to have access to their customer database, which they accessed in order to steal records and information about the specific clients of the company. This information included customer credentials, which could easily be used for targeted spear phishing attacks. InfoArmor identified multiple such instances where cybercriminals had infiltrated electronic health record (EHR) systems and leveraged the compromised data for extortion and ransom (InfoArmor 2016).

For the protection from all the above potential threats, businesses must monitor the cybecriminal underground and react to potential darknet threats and stolen assets by involving a range of departments, from IT, legal, HR and marketing, while focusing on the following key points (Delamarter 2016):

- Apply new, up-to-date and strong encryption methods on all sensitive corporate data.
- Set up state-of-the-art intrusion detection systems (IDS) and intrusion prevention systems (IPS).
- Monitor darknets to identify trends in term of attacks and newly discovered vulnerabilities.
- Monitor organisation's hardware and network use and investigate darknet access.
- Prepare a response plan to guide the corporate response to cyberattacks, loss of sensitive data or IP appearing on the darknet.

2.2.1.5 Law Enforcement Agencies

Law enforcement agencies constitute a third category of users acting as an umbrella to protect legitimate users from one side and fight criminal users on the other. Their main tasks and roles involved are described in this chapter.

The main tasks LEAs carry out not only in the physical world but also in the surface and the Dark Web environment, applying different techniques and tools for each environment, include monitoring, surveillance, disruption, investigation, enforcement and data intelligence gathering, processing and analysis for crime prevention, investigation and prosecution purposes based on evidence.

Behavioural and technical elements characterising crime on the Dark Web and markets can be exploited for successful policing operations. The fact that criminals make mistakes focusing their efforts on hiding their identity on the Dark Web, but at the same time leaving traces on their real-world identity, enables LEAs to recover these mistakes through the observation of behavioural patterns or technical faults. Criminals also follow business strategies and patterns. Keeping record of the history of online profiles, although anonymous, gives the opportunity to track, categorise and profile customer behaviours possibly related to cybercrime activities and money flow-related financial transactions.

Regarding the investigation tasks in Dark Web and darknet market policing, there are three investigation phases which have been identified during the workshop in Dark Web, held as part of the European research project MEDI@4SEC, on September 26, 2017, held in The Hague, Netherlands:

1. The strategic investigation
2. The identification of suspects
3. The prosecution of suspects

How successful the first strategic phase of Dark Web policing will be depends either on intrinsic characteristics of the technology and environment of the dark markets or on organisational and governance aspects of the broader ecosystem where law enforcement needs to operate.

In the second phase of the identification of suspects, the goal can be that of the surveillance of suspects to research their identity and responsibility in the crime. If a criminal investigation leads to a geographical area, for example, when a server hosting dark market services is located, a local operation is typically set up. Finally, in the third phase of the suspect's prosecution, once an investigation successfully leads to an arrest, sufficiently solid proof and evidence need to be in place to bring the case to court.

Some techniques exploited by criminals such as misinformation, fake news, the adoption of undercover initiatives and the use of crypto could be leveraged by LEAs as well. Also, coordinating operations and stimulating a collaborative sharing culture between LEAs at local, regional, national, European and international level, reducing double efforts by sharing a common repository of data, information, knowledge, tools, practices and methods and putting a common legislation framework in place, were some of the main recommended actions that came out from the workshop as outcomes for the project's reports (MEDIA4SEC-TNO (Serena Oggero) 2017).

2.2.2 *Criminal Actors*

Another group of users can be added to these groups, namely, the criminals. Different sorts of criminal activities and services are exploited on the Dark Web including criminal trade, child sex abuse (CSA) and criminal services, such as murder for hire, human trafficking and hacking services (Biryukov et al. 2014; McCoy et al. 2008). Other 'grey' services, not necessarily illegal, can be exploited for illegal purposes, such as financial transactions that may facilitate money laundering; the distribution of informative material, typically dealing with illicit content as the 'making of a bomb'; and several fora and chat services often used for communications that can facilitate the growth of violent fundamentalism and even extremism.

2.2.2.1 Cybercriminals

Among the many situations that have contributed to the increase in hacking and cybercrime is the amount of information being passed and the overall dependency on the Internet and digital devices. Over the last decade, the number of financial transactions online has increased, creating a tempting target for cybercriminals. The Dark Web nowadays is an ideal place for cybercriminals, and that is because of the anonymity that it provides them. Cybercriminals take advantage of it and have turned almost all their activities in the dark markets. They use mostly the Dark Web to sell and transfer illicit goods and materials. For this taxonomy the term 'cybercriminals' is adopted for a variety of cybercrime stakeholders in order to depict traditional crimes through the use of computer systems (e.g. drug and firearm dealers, production and distribution of child abuse material, financial fraud, human trafficking, etc.). The Dark Web will continue to be the ideal environment for cybercriminals as demand and offer with this trading and anonymity method are more widespread than ever.

The Dark Web (Rowley 2017) is a safe place for cybercriminals and their aiming profit. Dark Web uses include buying and selling illegal goods such as drugs, stolen information, weapons, and malware, among others (Richard 2017). In addition, anyone participating as a member on forums, chat rooms and Dark Web sites can find a tonne of information and knowledge for committing crimes ranging from physical and online theft, to advanced hacking skills, to trading stolen IDs and passports, buying and selling drugs, weapons, malwares, etc. The Dark Web's marketplaces offer all the above illicit goods and services, and the payments are being completed with cryptocurrencies, like Bitcoin. In the hidden world of dark markets, the most popular illicit goods are drugs, while in the last few years there has been an explosion of child sexual abuse sites.

The Dark Web is a great place for cybercriminals because apart from the illegal trade and transactions the Dark Web promotes the sharing of best practices between them. Regardless of the differences in languages, skills, locations, etc., cybercriminals tend to communicate with each other, in order to exchange opinions, skills and information or even cooperate on a criminal level (Lefkowitz 2017a). Furthermore, the Dark Web has many illicit marketplaces that enable cybercriminals to monetise the crimes they commit, facilitating this way the underground economy. The more important illicit goods that cybercriminals trade are corporate data theft, credit card fraud, corporate insider threat such as valuable and critical data and information, emerging malware and emerging fraud techniques, drugs and child sexual exploitation.

Today, the list of things that cybercriminals dispose on the Dark Web markets has increased greatly (Chickowski 2016). There is a variety of almost all illicit goods you can imagine such as social security numbers, stolen credit card numbers, full identity information, drugs, paedophile images and videos and even human organs. The illicit shops and salespeople on the Dark Web literally operate as true businesses, offering even discount packages if someone passes a particular order amount. Generally, cybercriminals tend to use the Dark Web more and more in order to gain profit by trading illicit goods and services and material of child sexual exploitation and support or organise terrorist attacks.

Cybercriminals remain the most active group in cyberspace; their motive is to earn as much money as they can in order to fund their criminal activities (ENISA 2018). To achieve that, and taking into consideration the rising technology, cybercriminals have employed more sophisticated methods, like malwares and ransomwares. Through these, they are able to employ cyberattacks to victims with monetisation potential and sensitive, valuable data. If the cyberattack is successful and cybercriminals have taken access of crucial data, they extort their victims for not making the information and data publicly available. At this direction, the consistent increase in data breaches for a consecutive year is a clear indication for the interest of cybercriminals in monetising stolen information.

2.2.2.2 Hacktivists

Hacktivism, one of the digital forms of activism, involves employing hacking skills and tools in order to attack governmental institutions and private organisations, on most occasions. Hacktivists are working in groups motivated by sociopolitical beliefs and ideology. Furthermore, they are acting anonymously, and in most cases instead of engaging in healthy debates and sharing their ideas, they will be more aggressive using criticism (Sorell 2015). Hacktivists also use the Dark Web in order to buy or sell information that will aid them to serve their political and social ideologies (Sixgill 2018). Due to the fact that hacktivists try to harm their victims with the dissemination of sensitive information and breached data, they are turning to the Dark Web in order to assure that they will not be detected by the LEAs.

Usually, hacktivists share sensitive databases that contain lists of companies' employees, their full ID, phone numbers and email addresses. In that way, hacktivists serve their purposes, while at the same time they are appreciated from the dark forum members. In addition, there are plenty of forums consisting of non-hacktivist members, but the information that is posted and sold there is attractive for hacktivists who are willing to buy it and harm their enemies. Besides, many ideological hackers act on behalf of, or inspired by, terrorists employing low-level cyber tactics against opportunistic targets, such as website defacements, malicious code injection, DDoS attacks and socially engineered account takeovers.

Moreover, there is another form of hacktivism that is observed on the Dark Web and Internet generally. For instance, the most known hacktivist group is Anonymous (Vijay 2017). Anonymous is well known for its march protests in the physical world and for DDoS attacks and hacking in the digital world. Last year, Anonymous locked down thousands of Dark Web websites accessible through Tor in order to protest against child sexual exploitation material and drug selling that was moved through these websites. Anonymous justified their actions by claiming that the majority of data that was compromised contained child sexual exploitation material. After that, they offered to sell the compromised data back to the websites owners.

Anonymous launched a 'school' for hacktivists on the Dark Web, called OnionIRC (Zorabedian 2016), in order to train and share hacking technical skills with the members and the awareness of anonymity software. OnionIRC is a chat forum that anyone can join, enjoying the anonymity that encrypted communication provides. The forum was advertised to the public through a video posted to YouTube that hacktivists promoted through their Twitter accounts, and so far plenty of campaigns took place varying from DoS attacks, hacking email accounts and dumping documents online (doxxing) to advanced attacks that have compromised government agencies and other high-profile targets. Finally, the forum members generate comprehensible documentation and instructions on how to create your own Tor hidden services, chat networks and other technical stuff, and as a result, many young people who admire technology are charmed and attracted to these groups.

2.2.2.3 Black Hat Hackers and Virus-Hacking Tool Coders

Hackers, either black hat, white hat or grey hat, are using almost the same tools and techniques, but with different motives and goals. Black hats are elite hackers undergoing illegal activities. Even though, other actors in this chapter can be characterised as black hats (e.g. hacktivists), for this taxonomy, we identify as black hats individuals or groups with excellent computer skills (elites). Their primary motive is to earn money (e.g. hacking as a service) and on occasions to cause significant damages (e.g. destroy/steal confidential data) (Sabillion et al. 2016; Martin 2017).

To the same direction are oriented the virus-hacking tool coders which are individuals or teams of expert programmers, elite hacking tool coders with excellent computer skills as well. The main focus of these actors is to develop computer viruses/malwares/rootkits/exploits (malicious code) and hacking toolkits in order to distribute it either to earn money in black markets or freely. The main 'buyers' are non-expert individuals who want to become hackers (e.g. script kiddies, Sabillon et al. 2016).

Combined with the desire for gaining profit, the above categories may create or contribute actively to the design and implementation of various malware forms, the most well-known and profitable to be at the time, ransomware and DDoS attacks. The main purpose of the above is to panic their victims, creating the unavailability to their services and data, which is able to create a lot of damage and malfunction, especially to businesses and large companies. The victim, in order to gain access again to his computer or network system, is required to pay a large amount of money to the hackers. In other words, this is a cyber-extortion by which the hackers gain a lot of profit, and it is concerned to be a more and more increasing phenomenon. In addition, many hackers that do not intend to perform their own cyberattacks to websites and network infrastructures dispose of their malware by selling it for large amounts on the Dark Web, avoiding the risk of losing their anonymity and getting caught somehow by the law enforcement.

2.2.2.4 State-Sponsored Attackers

Attackers sponsored and driven by countries, cyber syndicates and cyber-terrorists, for various reasons in both times of war and peace. The attackers aim to cause damage by gaining illegal access to state and trade secrets, technology concepts, ideas and plans and in general artefacts of value for a country or state. Their intentions often include the harm and damages on critical infrastructure, and in general they seek to damage the state's economy (Rasmussen 2014).

With the rising of the Dark Web and the dark markets, cyberattacks are becoming more complex and intense than ever, posing a significant danger and concern to governments. For that reason, one of the governments' reactions is the development of highly sophisticated malwares (Damien n.d.) in order to exploit zero-day vulnerabilities and compromise the target. However, such actions pose new concerns just in case a government's malware would be compromised and become available for anyone that would pay for it at on the Dark Web. This is not far from reality. It has already happened and you could imagine how cybercriminals, cyberterrorists and even state-sponsored hackers would enhance and use it in unpredictable ways, like harm critical infrastructures.

In addition, in 2013, hundreds of millions of Yahoo email accounts (Larson 2016) were compromised, and their data (such as phone numbers, passwords, security questions and alternative emails) were breached. All of this data had been available since then on the Dark Web, to any enterprise or entity that desired this data thesaurus with the money equivalence up to $300,000. Between this data breach, there were data concerning employees of agencies including the FBI, NSA, the White House and officials in the UK. Yahoo attributed that attack, which compromised 500 million accounts, to a 'state-sponsored actor' without condemning a specific country. Generally, nation states and their criminal partners can disrupt in many ways commerce and cyber defence. Their prime targets are everyday businesses of all types and sizes for potential cyberattacks which will lead to data breaches (IDExperts 2015). Cyber-espionage is a cheap and effective way to get the data you want plus cause economic and political damage. Also, it can be performed in order to embarrass a foreign government.

Cyber-espionage is widely applicable by industrialised nations which compete for world dominance in economic markets, in order to gain competitive advantage. For instance, a report revealed that China-sponsored hackers gained access to networks at many American companies and stole thousands of emails and documents concerning trade cases and more significant assets, such as designs for nuclear power plants. State-sponsored attackers use exactly the same methods as any other kind of cyberattack. However, cyber-espionage attacks are more likely to be multistage and persistent as in that case cybercriminals seek information and data that they can sell in a high price, such as credit card numbers or medical records, or compromise that their publicity will cause embarrassment to the target.

2.2.2.5 *Trade and Service Providers and Buyers*

Illegal trades and sells are growing exponentially although big dark markets (Bugge 2017) such as AlphaBay have been shut down by LEAs. This happens due to the fact that illicit goods have big demand and criminals seek for new clients online aiming at the biggest profit. The financial exchanges take place with cryptocurrencies such as Bitcoin. Most illegal trades that take place concern guns, explosives, drugs and stolen items such as credit cards, etc. Although, many dark markets such as AlphaBay and Hansa are currently shut down by the LEAs, Dark Web criminals are not being discouraged, but they are getting reorganised very quickly and continue their mission through new dark websites.

Almost any type of illegal and legally questionable products and services can be found on the Dark Web (Lee 2014). One of the most vivid examples of a well-known trading dark website was the Silk Road which is not active at the moment. The Silk Road was something like the eBay but most famous for offering drugs but also fake IDs and passports, guns and stolen credit cards. All the Silk Road enjoyed the anonymity that the Dark Web and Bitcoin provided them for some years, but eventually LEAs managed to trace and arrest the administrator of Silk Road which was shut down. After a while, a new similar dark market showed up, called Silk Road 2, though it had the same luck with Silk Road. Nowadays, there are a lot of active dark markets that are reported with detail in the url: https://darkwebnews.com/dark-web-market-list/.

The Dark Web seems to be again the solution for the drug dealers that sell through the Dark Web huge amounts of drugs. The anonymity and complexity of the Dark Web created the right conditions for the sale of dangerous synthetic drugs (Popper 2017), such as fentanyl, which reach to the consumer via mail after the deposit of the necessary Bitcoin or other cryptocurrency amount.

Synthetic drugs are far more addictive and overtaking than the 'classic' drugs such as heroin or cocaine. Also, a synthetic drug's particular dose can be fatal compared to the same dose of heroin. Another crucial point is that the classic drugs have quite volume and are parcelled so it can be detected during their delivery; synthetic drugs from the other side can easily fit in a mail envelope and can pass without being detected. Although Silk Road, the first big Dark Web drug market, was shut down, other similar markets emerged. Through these markets the classic drugs that are sold concern the minority of overall traffic; however the sale of synthetic drugs is increasing more and more, as well as the number of deaths that are being caused by overdoses.

Below is a list of 'products' and 'services' offered in the principal black markets:

- DDoS attacks (ordered or carried out for the purpose of extortion)
- Theft of personal information and data to access e-money (for the purpose of resale or money theft)
- Theft of money from the accounts of banks or other organisations
- Domestic or corporate espionage
- Blocking access to data on the infected computer for the purpose of extortion
- Trojans
- Exploits and exploit bundles
- Rootkits
- Crypters
- Fake documents and IDs
- Stolen credit card and other credentials
- Dedicated-server-hosting services
- Proxy-server-hosting services
- VPN services
- Pay-per-install services
- Denial-of-service attack services
- Spamming services
- Flooding services
- Malware checking against security software services
- Social-engineering and account-Hacking Services
- Malware (Trojans, bank stealers and backdoors)

- Drugs
- Crypting services
- Compromised hosts
- Remote access tools/Trojans (RATs)

2.2.2.6 *Insider Threat*

Insider threat can cause monetary losses to an organisation and are the results of actions or errors caused by individuals within the organisation. As presented in the 15th annual CSI Computer Crime and Security Survey reports, there are two separate threat vectors contributing to insider threats: firstly, employees with malicious intents against the organisation they working for (e.g. leak/sale non-public information, data breaches, etc.), and secondly, employees within the organisation who have made some kind of unintentional blunder. The report reveals that the majority of losses are due to non-malicious actors (Richardson 2010/2011).

Organisations should be extra careful by protecting themselves from external threats (Lefkowitz 2017b). They face many threats both in the physical and digital world, not only from externals but also from their employees. Many organisations are focused on the external threats because they may not have the time or expertise to identify insider threats. This is a crucial case, as there is possibility for the organisation's prestige to be significantly harmed and damaged due to a data exposure of an insider threat.

Insider threats are implemented when employees or disgruntled employees get access, authorised or not, to organisation's systems and data. Insider threat attacks may be prepared for a long time in an obscure way, and for this reason, it is difficult to target an inside threat that does not have a suspicious attitude, though they are focused to their plan and very dangerous. Nowadays, organisations have finally taken into consideration insider threats and have adjusted to better security politics and measures, as many criminal schemes are possible to come true only with an insider's help. The recruitment of the insiders takes place through the anonymised Dark Web. There are more than a few cases whereby organisations have found data and information in forums posted by insider.

The common practice of organisations is to check the history and the background of their employees; however many concerning clues of their character are not identified from the screening procedure. So, it is relatively easy for actors from the Dark Web to approach them and 'cooperate' by harming the employee's organisation. Generally, actors recruit insiders in order to take access to confidential information, or from their own insiders, they are trying to sell this information on to anyone who is interested in this. Nowadays, organisations and enterprises pay too much attention to the external threats, due to the extended cyberattacks, ransomwares, etc. (Kitten 2016). However, there is always one threat that should be taken into serious consideration as it often takes place and its impact is usually severe. Like all the other illegal activities, the Dark Web is an appropriate place to reach insiders. Insiders usually sell their login credentials or by being recruited expose personal data or IP from their organisation.

Insider threats made their appearance to the Dark Web too (Metzger 2017). Beyond the aforementioned anonymity of the Dark Web, financial and personal data have a great value. Insiders can be a deadly, irreversible threat to an organisation. With rising technology and the fall in living standards, this phenomenon knows unpresented raise. The whole situation about insider threat has changed so dramatically, because a few years ago the insiders had not had the same profit. In addition, stealing the data was way more difficult because most of them had been stored in physical ways (paper, files and protected rooms) and in case of theft distribution to the malicious employers was not the safest. The evolution of technology combined with the digitalisation of organisations' archives, the possibility of transmitting them with encrypted communication applications and Dark Web secrecy was a significant factor to the widespread ofinsider threat.

Statistical research indicates that nowadays people who are working for organisations, businesses, etc. that possess critical data are more tempted than ever to sell them and harm the organisation. This comes from the fact that more and more people have financial problems, and selling critical data and similar illegal actions are a way of ensuring a satisfying amount to cover their needs. Generally, the insiders can be classified in three categories: the malicious ones who just want any reason to harm an organisation, the employees that expose their companies by mistake or because they ignore the necessary security and the compromised, who are being coerced into doing so. In the last category, there are good and bad, and it depends on the pressure they get or their personal situation (financial, psychological) whether they will harm their organisation or not.

Insiders are being actively recruited by criminals operating on the Dark Web (Litan 2016). They are usually disgruntled employees that were fired, and they are seeking firstly to harm and revenge on employers and then to earn money by their illegal actions and by selling their insider knowledge and services to bad guys on the Dark Web.

2.2.2.7 Cyberterrorists

Terrorist groups are increasingly using the Dark Web to recruit and train new members, share information and organise attacks in the real world. Furthermore, terrorist organisations are using the anonymity and security of the Dark Web to disseminate training guidelines for cyberattacks, to less experience supporters (Sieber and Brunst 2007).

Terrorists are increasingly using encrypted applications and the Dark Web to create a digital safe haven for communications, logistics and financing of their activities. Currently, the most active actors on the Dark Web are cybercriminals generally and (cyber)terrorists (Office of Homeland Security and Prepardeness 2018). Cyberterrorism is a modern form of terrorism that combines the two biggest fears of our times: cyberspace and terrorism. The Internet and more specifically the Dark Web are a perfect place for cyberterrorists as they provide encrypted and quick communication between terrorists and eliminate all the distances. Equally active on the Dark Web is the propaganda and extremism by terror groups such as ISIS.

Cyberterrorism (ENISA 2018) so far is not an emerging and highly concerning threat because cyber capabilities of terrorists are low. The most serious 'attacking' activities of terrorists concerning the cyberspace are mainly hacking actions, such as defacements and DDoS attacks. Furthermore, it seems that cyberterrorists have turned their attention in developing capabilities for cryptocurrencies, as it is the best and anonymised source for their funding and their trading on the Dark Web, vital for the completion of their operations and to equip them with the necessary equipment. Moreover, cyberterrorists may be interested in using dark markets to purchase cybercrime services, such as malwares and ransomwares in order to perform cyberattacks that will as a result extort and gain financial profit.

Generally, the idea is that terrorism generates fear of a massive, destructive, physical attack resulting in the loss of human. A cyberattack, like the DDoS attack itself, does not cause this kind of fear or destruction; however analysing carefully the potentials, it could be just the beginning of a chain of events. For instance, if through a massive cyberterrorist attack telecommunications and first response services would be disabled, the effects probably would be catastrophic and their avoidance inevitable. So, an attack of this kind could easily be combined with a physical attack (bombing, slaughter, etc.) at a big area where the communication systems would be paralysed and the coordination and the tackle of attack would be a major issue.

Cyberterrorist attacks are totally familiar with the common cyberattacks with the difference that cyberterrorists would perform them where it would have the potential to cause more damage, such as a power grid. The purpose is different also. Cybercriminals desire to gain mainly financial profit from an individual or an organisation through a cyberattack, while cyberterrorists desire to cause fear to people and organisations through a coordinated cyberattack, promoting at the same time their sociopolitical agenda.

The current situation is how to confront all these Dark Web activities from terrorist groups, such as ISIS, which turn to become more and more experienced in cyberspace (EZPC Indy Admin 2018) and try to recruit more cyberterrorists through radicalising messages, videos and extremist content. Undisputedly, the safety and anonymity that the Dark Web provides to terrorists contributed actively to the widespread of propaganda to their sympathisers across the globe. Already, a lot of efforts are taking place by cyberterrorists to harm and terrorise people and organisations that are against their ideology and present critical infrastructure and information. It is going to be a challenging situation to tackle and confront the cyberterrorists that tend to be more experienced in cyberspace under the safety and their presence on the Dark Web.

2.3 Concluding Remarks

The Dark Web's major features and characteristics such as anonymity and security provide the same advantages to two groups of users, the legitimate and the criminals, to act differently and oppositely in the same community space.

The Dark Web is not only all about illicit and criminal activities as it is usually connected to by the majority of people. Ordinary people, journalists, activists, business executives, IT professionals and LEAs are some of the main legitimate actors of the darknet.

With the indication that privacy, freedom of speech, truth and trust of news and sensitive information are either unavailable or under control, surveillance, monitoring and attacks in our day to day lives allow actors to go under the surface and use the Dark Web infrastructure. Many media groups and human rights organisations in the role of activists already use and recommend the use of darknets, whereas IT professionals and business executives also use such infrastructures for security, testing and providing Internet access with confidentiality during network failure cases that may occur on the normal net.

References

A. Bugge, *Dark Web Drug Market Growing Rapidly in Europe: Report* (2017)

A. Biryukov, I, Pustogarov, F. Thill, RP. Weinmann, Content and popularity analysis of Tor hidden services. *In 2014 IEEE 34th International Conference on Distributed Computing Systems Workshops (ICDCSW)* (pp. 188–193). IEEE. (2014)

E. Chickowski, *Cybercrime: A Black Market Price List From The Dark Web* (2016)

CIVIL, *A Decentralised Marketplace for Sustainable Journalism-How Does It Work.* [Online] (2018), Available at: https://joincivil.com/how-it-works/. Accessed 19 May 2018

Con, *Can Journalism be Bound for the Dark Web?* [Online] (2018), Available at: https://darkwebnews.com/dark-web/can-journalism-be-bound-for-the-dark-web/. Accessed 19 May 2018

Damien, *Researchers Discovered a Government-Made Malware on the Deep Web* (n.d.)

A. Delamarter, *The Darknet: A Quick Introduction for Business Leaders.* [Online] (2016), Available at: https://hbr.org/2016/12/the-darknet-a-quick-introduction-for-business-leaders. Accessed 19 May 2018

ENISA, *ENISA Threat Landscape Report 2017–15 Top Cyber-Threats and Trends* (s.l., s.n., 2018)

EZPC Indy Admin, *A Close Examination of Cyberterrorism* (2018)

I. Glick, *Darknet: Where Your Stolen Identity Goes to Live.* [Online] (2016), Available at: https://www.darkreading.com/endpoint/darknet-where-your-stolen-identity-goes-to-live/a/d-id/1326679. Accessed 19 May 2018

IDExperts, *State Actors and Cyber-Espionage: Computer Warriors vs. Your Business* (2015)

InfoArmor, *AlphaBay Market: Pilotfish Technology Source Codes are for Sale in the Underground.* [Online] (2016), Available at: https://blog.infoarmor.com/news/alphabay-market-pilotfish-technology-source-codes-are-for-sale-in-the-underground. Accessed 19 May 2018

Intelliagg, *DEEPLIGHT: Shining a Light on the Dark Web* (s.l., s.n., 2016)

T. Kitten, *How the Dark Web Presents New Insider Threats* (2016)

S. Larson, *Hackers are Selling Yahoo Data on the Dark Web* (2016)

T.B. Lee, *The Dark Web: What It Is, How It Works, and Why It's Not Going Away* (2014)

J. Lefkowitz, *How Cybercriminals Use the Deep and Dark Web to Target Financial Organisations* (2017a)

J. Lefkowitz, *Securing an Organisation Against Insider Threats* (2017b)
A. Litan, *Insider Threats Escalate and Thrive in the Dark Web* (2016)
D. McCoy, K. Bauer, D. Grunwald, T. Kohno, D. Sicker. Shining Light in Dark Places: Understanding the Tor Network. In: Borisov N., Goldberg I. (eds) Privacy Enhancing Technologies. PETS 2008. Lecture Notes in Computer Science, vol 5134. Springer, Berlin, Heidelberg. (2008) https://doi.org/10.1007/978-3-540-70630-4_5
MEDIA4SEC-TNO(Serena Oggero), *Workshop 3 Report: Policing the Dark Web.* [Online] (2017), Available at: http://media4sec.eu/downloads/d2-5.pdf. Accessed 19 May 2018
M. Metzger, *InfoSec 2017: Dark Web and Economic Downturns Fueling Insider Threats* (2017)
A. Murray, *The Dark Web is Not Just for Paedophiles, Drug Dealers and Terrorists.* [Online] (2014), Available at: https://www.independent.co.uk/voices/comment/the-dark-web-is-not-just-for-paedophiles-drug-dealers-and-terrorists-9920667.html. Accessed 19 May 2018
Office of Homeland Security and Prepardeness, S. o. N. J., *2018 Threat Assessment Series: Enemies in Cyberspace – Cyberterrorism and Terrorists' Use of the Internet.* [Sound Recording] (2018)
N. Popper, *Opioid Dealers Embrace the Dark Web to Send Deadly Drugs by Mail* (2017)
N. Rasmussen, Cyber Security, Terrorism, and Beyond: Addressing Evolving Threats to the Homeland (2014). Retrieved from https://www.dni.gov/files/NCTC/documents/news_documents/cyber_security_terrorism_and_beyond.pdf
RFS, *Online survival Kit | Reporters without borders.* [Online] (2018), Available at: https://rsf.org/en/online-survival-kit. Accessed 16 Apr 2018
U. Sieber and P. Brunst. Cyberterrorism and Other Use of the Internet for Terrorist Purposes: Threat Analysis and Evaluation of International Conventions. In Council of Europe (Ed.), *Cyberterrorism – the use of the Internet for terrorist purposes* (pp. 9–105). Strasbourg: Council of Europe Publishing (2007)
Richard, *Is Dark Web a Hidden Place For Hackers, Cybercriminals and Whistleblowers?* (2017)
R. Richardson, *15th Annual Computer Crime and Security Survey* (CSI – Computer Security Institute, s.l., 2010/2011)
O. Rowley, *Shedding Light on the Deep & Dark Web: Bringing Risk Intelligence to Bear for Business Benefit* (2017)
R. Sabillon, V. Cavaller, J. Cano, J. Serra-Ruiz. Cybercriminals, cyberattacks and cybercrime. *In 2016 IEEE International Conference on Cybercrime and Computer Forensic (ICCCF)* (pp. 1–9). IEEE . (2016, June)
Sixgill, *Dark Web Sites as a Platform for Hacktivist Warfare* (2018)

TOR, *Users of TOR*. [Online] (2018), Available at: https://www.torproject.org/about/torusers.html.en. Accessed 19 May 2018

Vijay, *Hacktivist Group Anonymous Takes Down Nearly 10,000 Tor Websites on Dark Web* (2017)

J. Zorabedian, *Anonymous Launches OnionIRC – A School for Hacktivists on the Dark Web* (2016)

T. Sorell. Human rights and hacktivism: the cases of Wikileaks and anonymous, *Journal of Human Rights Practice, 7(3),* pp. 391–410. (2015)

C.D. Martin. TAKING THE HIGH ROAD White hat, black hat: the ethics of cybersecurity, *ACM Inroads, 8*(1), pp. 33–35. (2017)

CHAPTER 3

Terrorist Activities in the Dark and the Surface Web

Euthimios Lissaris, Georgios Giataganas, Dimitrios Kavallieros, Dimitrios Myttas, and Emmanouil Kermitsis

3.1 Introduction

Terrorists, and especially jihadists, the last 20 and more years are a significant and constant menace mainly for the African and European countries. As anyone could see over the last few years, more and more terrorist attacks take place in countries such as France, Spain, the UK, Germany, Iraq, Afghanistan, Syria, Nigeria and Somalia. This makes terrorism

E. Lissaris • G. Giataganas • D. Myttas • E. Kermitsis
Center for Security Studies-KEMEA, Athens, Greece
e-mail: e.lissaris@kemea-research.gr; g.giataganas@kemea-research.gr; d.myttas@kemea-research.gr; e.kermitsis@kemea-research.gr

D. Kavallieros (✉)
Center for Security Studies-KEMEA, Athens, Greece

University of Peloponnese-Department of Informatics and Telecommunications, Tripoli, Greece
e-mail: d.kavallieros@kemea-research.gr

B. Akhgar et al. (eds.), *Dark Web Investigation*,
Security Informatics and Law Enforcement,
https://doi.org/10.1007/978-3-030-55343-2_3

nowadays, more than ever, a worldwide organised and persistent threat. But what makes terrorists such a significant, countable enemy for governments and law enforcement agencies? Terrorists, in order to fulfil their goals, must be able to make a public online "presence", while at the same moment they must remain undetected from law enforcement agencies. So far, terrorists spread their propaganda and post pictures, videos and other media concerning their actions on social media because of the extensive use by the majority of people. Nevertheless, the use of social media such as Facebook, Twitter, YouTube, etc. has resulted in the cooperation of law enforcement agencies with organisations such as Europol and Interpol in order to detect and identify terrorists from their online activity and to erase and deactivate their posts or their websites.

Terrorists continue to use the most known social media to spread their propaganda and to radicalise and recruit people, mainly young members of vulnerable and marginalised society groups, filled with sentiments of injustice, exclusion or humiliation, although, lately they have found a different way to communicate with other members of the terrorist group, other terrorist groups, their supporters and their potential recruits. This way is through the use of the Dark Web.

The Dark Web is a "secret", concealed part of the Internet that is used from many users that desire to maintain their anonymity and stay undetected. So, terrorists nowadays mainly use the Dark Web in order to communicate with each other and to share information and training material with their supporters or new recruits on how to make bombs, for attack plans and other useful things that are part of explicit terrorist tutorials. Terrorists also use encrypted platforms and apps such as Telegram, WhatsApp, Surespot and other end-to-end encrypted tools in order to avoid their identification from law enforcement agencies, their geographical location and the eavesdropping of their conversations. These apps give terrorists a lot of "freedom", and they are not worried at all about how to make their conversations and their transactions safely and secure without being traced.

Finally, terrorists use the Dark Web for their ultimate goal: the group's survival. In order to be able to fulfil their goals and make their attacks, terrorists need a constant and increasing flux of money for funding the necessary equipment and their missions. The Dark Web is the ideal place for this. Through their own Dark Web websites, terrorists are able to get funding from their supporters, sell goods such as oil and weapons and buy equipment and whatever they need in order to continue their plans and actions.

3.2 Terrorist's Activities on the Web

According to Europol, the investigations for terrorist actions in Europe have revealed that the use of the Internet is an integral component in any terrorist plot (Europol EC3 2017). Also, the importance of the Internet for the terrorists can be proved by the significant increase in terrorist websites. Over the last few years, these sites have increased exponentially to some thousands. Having in mind that the origins of all major jihadist terrorist groups are from regions located outside the EU, they were obliged to use the Internet to conduct their activities (Europol EC3 2016). The Internet has turned to be a powerful tool in the hands of terrorists, which they use to achieve their mission and fulfil their goals, such as recruiting new members, communication, propaganda, valuable information- and knowledge-sharing purposes, coordination of attack plans and raising of funds.

While seeking for anonymity as a way to avoid being traced by law enforcement agencies, they have turned to the Dark Web (Chertoff and Simon 2015). Although the Surface Web might attract more of an audience than the Dark Web, however the latter seems to be more suitable for recruitment, financing, planning and training.

3.2.1 Propaganda

Before the growth of the Internet, the publicity of terrorist's causes and activities depended on traditional media such as television, radio or newspapers; however, the multistage processes of editorial selection by these media created difficulties to public terrorists' propaganda. To overcome these difficulties, terrorists created their own websites. In this way, terrorists have direct control over their content, and they can easily address their messages to target audiences. As noted earlier, terrorists, which are located outside EU, use the Internet for online propaganda, in order to reach out to audiences in EU member states. In this way, terrorists intend to engage potentially vulnerable people, living in the EU, with the armed struggle that they conduct in their areas of operation (Chen et al. 2008). The Internet gave them further opportunities to shape their messages in ways that manipulate both their own and their enemy's image. As a result, nowadays, almost everyone has seen propaganda videos and photos published on terrorist sites or re-uploaded onto news websites. For example, the Al-Qaeda promoted, via the Internet, an armed jihad in order to achieve their goal of an Islamic caliphate. They formed a media arm, known as As-Sahab, with the purpose to spread their propaganda and ideology through videos and statements of their leaders (Stratfor 2016). Moreover, the Daesh created a massive media legion in order to manipulate the Internet and to spread the group's propaganda. According to Charlie Winter's "Documenting the Virtual 'Caliphate'", Daesh has under its central media command 7 media agencies and 37 media offices operating in different provinces (Quilliam Foundation 2015).

3.2.2 Propaganda and Social Media

Terrorists have worked and experimented with the social media in recent years, and the results persuade them to conduct their activities on the safe environment of the Internet. According to Europol's European Union Internet Referral Unit (EU IRU), there are over 150 social media platforms which are abused by terrorists in order to succeed in the dissemination of their propaganda.

The main social media platform that Islamic State (IS) is using to spread their ideology on the Internet is Twitter (Berrada and Boudier 2017). IS members are using various twitter accounts, "tweets" and messages regarding their faith, hate against non-Muslims and the necessity to attack back. IS has achieved an uninterrupted online presence on social media, through a "sophisticated" communication strategy which relies on a network of core supporters (core disseminators). Thus, they prepare their media campaigns in encrypted social media such as Telegram, and after that their message is spread to the wider social media network (Europol EC3 2016). Nowadays, with the territorial, infrastructure and human resource loss for the terrorist, there has been a noticeable decrease in the release of new audio-visual material, which is also followed by lower production of textual content and photo reports. In order to compensate, the terrorists have created special social media accounts (i.e. Telegram channels) operated by core disseminators and bots and dedicated to the regular re-uploading of older productions. To ensure a long period availability of content, this procedure is repeated to a large number of pro-IS channels and advertised without links to social media. Also, this procedure of re-uploading old, high-profile propaganda items helps terrorists to achieve two basic priorities. The first one is to maintain a permanent virtual presence on the Internet, and the second one is to leave this virtual content to the future generations as a legacy (Europol EC3 2016).

3.2.3 *Propaganda and Dark Web*

According to what was mentioned before, IS terrorists are using the Dark Web not only as a means to perform propaganda via videos, images and announcements but also as a means to communicate with each other. The communication is provided through smartphone applications, which are accessible for downloading only in Dark Web, like Orbot (Guardian Project 2016). The shift to the Dark Web was more explicit after the November 2015 Paris attacks (BBC News 2015). After the specific terrorist attacks, a propagandistic message was sent by a Telegram channel (Fig. 3.1) which communicated several links to a Tor hidden service with an ".onion" address. With this message terrorists announced a new "dark" website that was a reflection of Isdarat, a popular website that publishes ISIS materials on the clearnet (Site Staff 2015).

The exact content of the original message was:

> *Due to severe constraints imposed on the #Caliphate_Publications [Isdarat Releases] website, any new domain is deleted after being posted.*
>
> *We announce the launch of the website for "dark web."*
>
> **It will work for the TOR users and the normal users.*
>
> *Link for the TOR users: http://isdratetp4donyfy.onion*
>
> *Link for the normal users: http://isdratetp4donyfy.onion.link*
>
> *And we promise you that we are continuing to try to get a normal and new domain and we will post it Allah willing when it is obtained besides [as well as] the TOR domain.*

As we can see, Fig. 3.1 was viewed by 7629 users who followed the IS propaganda channel on the Telegram platform. Thus, IS, apart from the surface website, had successfully obtained a Dark Web version website in which the propaganda content was available at its .onion address (Fig. 3.2).

Over the past few years, the top-level domain name of the Isdarat website has undergone numerous changes. That happened either because cyberattacks successfully managed to locate it or because the domain was being reported to the hosting company (Krypt3ia 2015). The main goal of creating the "dark" version of their website was, on the one hand, to

قناة إصدارات الخلافة

بسم الله الرحمن الرحيم
نظراً للتضييق الشديد على موقع #إصدارات_الخلافة
بحيث أنه يتم حذف أي نطاق جديد بعد نشره

نعلن إنطلاق الموقع على "Dark web"
*وسيعمل لُستخدمي الTor وللمستخدمين العاديين

رابط مستخدمي الTor :
http://isdratetp4donyfy.onion

رابط المستخدمين العاديين :
http://isdratetp4donyfy.onion.link

ونعدكم بأننا مستمرون فى مُحاولة الحصول على نطاق جديد عادي وسننشره إن شاء الله عند الحصول عليه بجانب نطاق الTor

{ولله العزة ولرسوله وللمؤمنين} 7629 2:05 PM

Fig. 3.1 Screenshot of the original Telegram message (INSITE Blog on Terrorism and Extremism 2015)

secure their identity from the LEAs and, on the other hand, to preserve the safety of the website from takedown operations. There are also benefits for the visitors, because they necessarily have to use Tor in order to access it, which masks their IP address and protects their identity (Fig. 3.3).

The website also contained translations from various statements issued by Daesh in English, Turkish and Russian. Even though the website is hosted as a Tor service, the propaganda media was provided by surface websites such as Google video. As a result, the Dark Web users are directed, through the URLs, to the surface websites in order to view such videos. Moreover, the hosts can easily remove all of these media from their websites. Furthermore, through the new website, the users were directed to the terrorist group's private messaging portal on Telegram, in order to accomplish their encrypted communication.

In an effort to enforce the online anonymity, an Al-Qaeda group distributed a manual entitled "Tor Browser Security Guidelines". The manual offered all the needed details of how someone could possibly download and install the Tor browser and, even more, how to delay geolocation and identification procedures by law enforcement agencies. It has also been

Fig. 3.2 Onion link for Dark Web propaganda site in Twitter (Israels ACG 2018)

noticed that some jihadi Twitter accounts had given out security advice for accessing the site by providing the link with an ".onion.link" suffix. Using the latter rather than the normal ".onion" suffix allows users to access the site without routing their traffic through the Tor network.

3.2.4 Content of Propaganda

It is undoubtful that propaganda which occurs by jihadists is constantly augmenting. In the last two decades, jihadist propaganda has advantaged its methods in order to reach the greatest influence on its audience. The Internet has significantly expanded the opportunities for terrorists to secure publicity (Chen et al. 2008). Now, in an effort to gain ground in the younger population, jihadist online propaganda tries to combine elements from both lifestyle and online gaming. It is very crucial for investigators to understand the propaganda themes and their motivations (Europol 2017).

In contrary to popular belief, most terrorist sites do not have the intention to advertise their violent activities. Instead, the main purpose is to

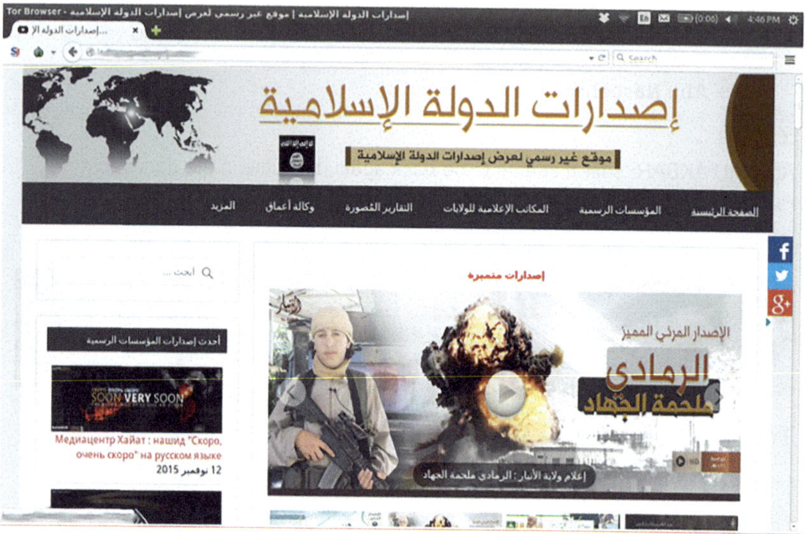

Fig. 3.3 Screenshot of Isdarat dark website (Krypt3ia 2015)

convince online audience about two issues: the restrictions placed on freedom of expression and the plight of comrades who are now political prisoners. Both of these issues are considered to create sympathy to the Western audiences which advocate the freedom of expression and reinforce the political opposition to be silent. In an effort to make their enemies feel anxious and shamed, they emphasised on the anti-democratic measures, which were occurring against them. The above terrorists' strategy is well suited for the Internet, which for its users is considered to be the symbol of free, unlimited and uncensored communication.

It can be easily said that jihadist propaganda varies over time, in order to accomplish the greatest impact at any given point. For instance, IS propaganda successfully convinced many that the group stands for victorious Islam. The purpose was to purport the defence of Islam and also the creation of an Islamic state. IS policies were the ones who supported this attempt, by promising justice and prosperity and providing services for the population. All of the above were shown by videos which emphasised on the implementation of social policies and religious education. The following year the content of IS propaganda was radically changed (Europol 2017). Moreover, Islam seemed to be under attack by a reputed alliance between the Westerns, Jewish and Shi'I, in contrast to a victorious one, embodied by IS. This was in fact one of the most considerable replacements that the IS propaganda had performed.

The IS leadership claims to follow what the sources of Islam dictate, in which in the day of judgement, Islam will dominate. Considering the above theory, IS leadership claims that its single battle against its hostile fronts happens due to the final battle. However, the *Dabiq* magazine (the *Dabiq* magazine name is referred to as a place in the northern countryside of Halab or Aleppo, where alleged events leading up to the end of time would take place), publishes in August 2016 that IS was fighting the West for no other reason than its "unbelief". Trying to deter any effort of dissociation and dispute in its region, IS proceeded to publish a great number of videos. These were attempted in order to show the individuals that were described as those who betrayed and even spied on IS leadership. Spies for the anti-IS coalition were murdered icily. Beheading and shooting from a close range seem to be the murder practices, which Muslims promoted. All are captured by multiple cameras. This can be recapitulated briefly that these videos used both repetition and the effect of slow motion. Thus, the impact of the murders was effectively enhanced. Referring to their supporters in the West, IS claimed that they had to retribute justice to whoever supported and fought for the international anti-IS alliance. IS also turned against Muslim scholars who opposed to them, likewise distinguished representatives of Sunni Muslims in Europe.

Moreover, in June 2016, IS decided to lead a media campaign, in areas under IS dominion, in order to eliminate access to information given by individuals. In this campaign a large number of videos were shown, in an effort to persuade people to destroy their satellite receivers. Once again IS propaganda was successfully achieved, as the failures that the military had were replaced by victories. They also managed to promote an ostensibly peaceful and orderly lifestyle in combination with an untroubled administration. As far as women are concerned, they were entirely manipulated by men, and, in fact, they were obliged to marry IS fighters. Obedience and submissiveness to their husbands were principles. IS publications tried to force women to marry men who already had up to four wives. This was an attempt to stimulate and enhance the sexual motive for IS fighters. Their children would become the future fighters and would unquestionably join the IS leadership. In an effort to secure and obtain the organisation longevity, IS spread the following message: IS affiliates should take the reins and fight, in case of defeat in Syria and Iraq. More specifically, Abu Bakr al-Baghdadi who is the IS leader joined the IS members in November 2016, to maintain their stability and be preserved solid against the attacks of the anti-IS coalition. He also called IS fighters in areas outside Iraq and Syria to continue the fight and asked IS sympathisers to move to these regions. In 2016, the most active IS affiliates outside Syria and Iraq with regard to video production in Libya, Egypt and Afghanistan (Europol 2017). The result of the propaganda was the activation of many terrorist groups around the world that had pledged their support or allegiance to Daesh (IntelCenter 2016). In addition, attacks in the name of Daesh have been recorded in Australia, Bangladesh, Belgium, Egypt, France, Malaysia, Indonesia, Turkey and the USA amongst others (Europol EC3 2016).

3.2.5 *Rhetorical Structures of Terrorist Websites*

In order to justify their belief on violence, terrorist sites have applied three rhetorical structures (Chen et al. 2008). The first one is the claim that terrorists have no choice other than to turn to violence, as they mainly presented to be the weak ones. Thus, the only opportunity for them to face an oppressive enemy is violence. Terrorist sites are not willing to mention their criminal actions, while, at the same time, the government's actions against them are characterised with terms such as "slaughter", "murder" and "genocide". The tactic in which the terrorist groups appeared as persecuted and their leaders seem to be target of assassination makes them look like small and weak, hunted down by a strong state.

The second rhetorical device is referred to the legality of the use of violence. In this rhetoric, one may have noticed that the responsibility for the violence is given from the terrorist to the adversary. The opponent is presented as an enemy without mercy, whom the members of the movement are obligated to combat, in order to assure their people's rights.

The last and third rhetorical structure is to use, on a large scale, the language of non-violence, as a result to deal with the terrorist's violent image. These organisations uphold the idea that their main goal is no other than a diplomatic settlement, which will be fulfilled through negotiation and international pressure on a repressive government.

3.2.6 *Crime as a Service*

An additional activity of IS terrorists on the Dark Web is the quest for "hiring" criminals who are willing to offer their criminal services for

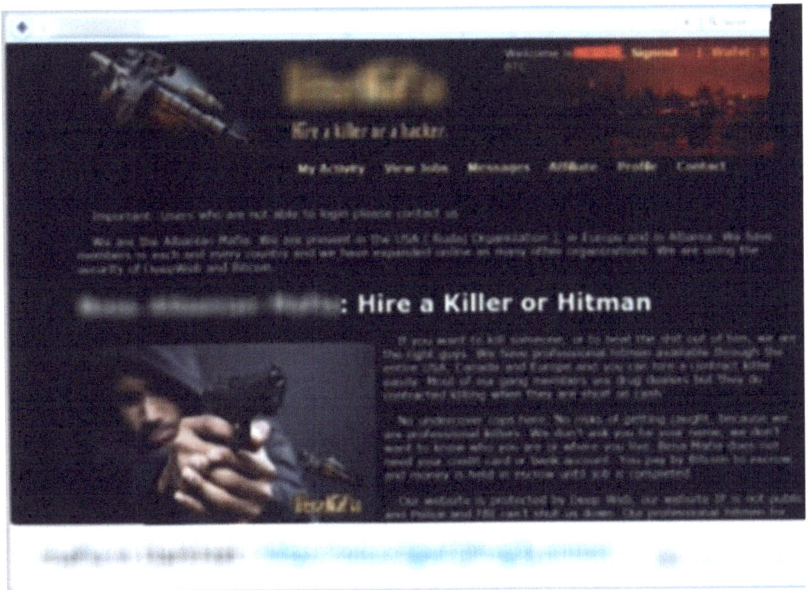

Fig. 3.4 Dark Web site for hiring criminals (Opinion 2017)

money. For instance, on the 14th of June 2016, an advertisement for a Dark Web site was shared via terrorist's Telegram channel, in which the services of Albanian assassins and hackers would be available. The site was accessible via Tor (MEMRI CJ LAB n.d.) (Fig. 3.4).

3.2.7 Communications

Over the last two decades, with the constant rise of technology, terrorists took the chance to change their strategy and evolve to high-tech and sophisticated groups. In addition, a worldwide network of hundreds of websites emerged that aimed to inspire, train, educate and recruit young people to engage in jihad against the Western countries. To achieve that, terrorists deployed two types of communications: secret and public. As far as public communication is concerned, online recruiting has exponentially increased in recent years by using Facebook, YouTube, Twitter and other types of social media that hundreds of millions of people use at a high grade on a daily basis. On the other hand, the Dark Web and encrypted

Fig. 3.5 Social media that terrorists use (Gate Stone Institute 2017)

communication are used amongst terrorists in order to remain undetected from the law enforcement and governmental agencies. For these reasons, big terrorist groups nowadays, such as ISIS, understand the importance of secure communication and have "embraced the (Dark) Web" more than ever (Gardner 2013) (Fig. 3.5).

3.2.7.1 Encrypted Communication and Dark Web

Terrorists use a lot of encryption methods and tools to conceal their communications from law enforcement and intelligence agencies, leaving them only the metadata to work with (Europol 2016). Apart from social media platforms, public or secret websites and tools, terrorists also use the Dark Web to achieve content concealment and anonymised communication with each other, anonymously (Bertrand 2015). The best disposable ways to achieve anonymisation and concealed communication are by using email, web chats, personal messaging hosted on Tor (The Onion Router) or with special tools that provide the users with anonymity and encrypted communication. More specifically (Finklea 2017):

- Email service providers: more often this only requires users to provide a username and a password to sign up. In most cases, it is totally optional if the user provides any other identity information or verification data, such as a phone number. In addition, email messaging is anonymous, and its storage is encrypted, protecting the communication's privacy. Alternative email services such as Hushmail and ProtonMail are widely used by terrorists.

- Web chat apps on Tor: e.g. Chat.onion (Chat Onion 2016) is an anonymous and fully encrypted peer-to-peer instant messenger using onion routing. Chat.onion uses onion routing (Tor) to send each message over several randomly selected proxy servers, hiding in this way information such as IP address, identity and metadata. In addition, its users are identified via 16 character checksums of their public keys. To connect with other users, the user must send them his ID, show them his QR code or use his camera to scan theirs (Fig. 3.6).
- Personal messaging with apps: e.g. Tor Messenger. It is a cross-platform chat program that aims to be secure by default and sends all of its traffic over Tor. It supports a wide variety of transport networks, including XMPP, IRC, Twitter, etc. Furthermore, it enables off-the-record (OTR) messaging and all secure updates automatically (Fig. 3.7).

There are more sophisticated technologies presented below containing encryption tools and anonymising software, increasing the difficulty of revealing the sender's identity or the content of message. These tools mask the unique IP address that identifies each device accessing the Internet and its location, reroute Internet communications via one or more servers to jurisdictions with lower levels of enforcement against terrorist activity and/or encrypt traffic data (Reisinger 2016).

- Bitmessage is a P2P communications protocol used to send encrypted messages to another person or to many subscribers. It is decentralised and trustless, which means that you do not have to trust any

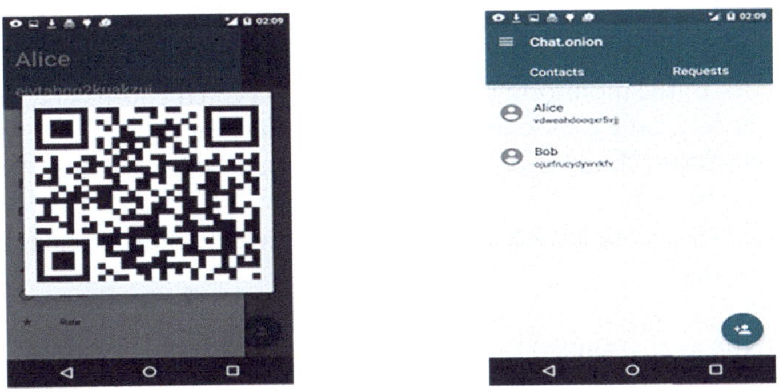

Fig. 3.6 Onion chat app (Chat Onion 2016)

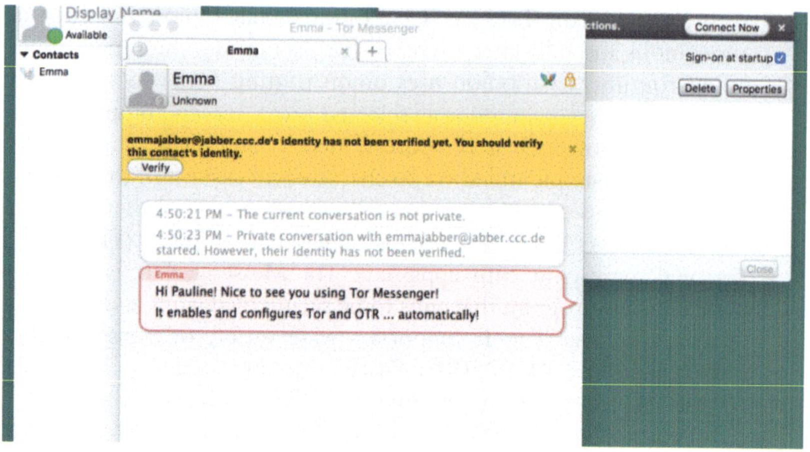

Fig. 3.7 Tor Messenger (Wired 2015)

entities. It uses strong authentication preventing the spoofing of a message's sender, and it aims to hide "non-content" data, like the sender and the receiver of the messages, from passive eavesdroppers like those running warrantless wiretapping programs (Fig. 3.8).

- Ricochet also uses the Tor network to reach user's contacts without relying on messaging servers. It creates a hidden service that has the ability to hide the user's location and IP address. In addition, the user gets a unique address that looks like ricochet:xxxxxxxxxx (letters and numbers). Other Ricochet users can use this address to send a contact. The user is able to check when his contacts are online in order to send them messages rapidly, with the updated features and files. Furthermore, contact lists are never exposed to servers or network traffic monitoring. Everything is encrypted end-to-end, and in that way, only the original recipient can decrypt it.
- VeraCrypt, TrueCrypt and Hardskat (for hard drive or flash drive encryption).
- MEGA or SpiderOak (for security of information or data stored on cloud).

Finally, the chat rooms on Tor are more than 50 at the moment, while there are plenty of Dark Web sites that host private messaging. There are already a lot of ISIS supporters that use these apps and tools securing their

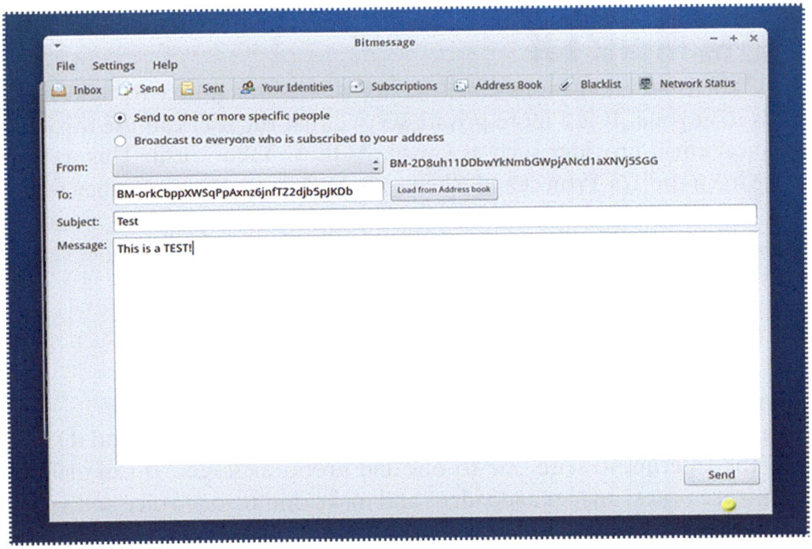

Fig. 3.8 Bitmessage (BBARCESAJ 2013)

virtual identities, their data on computers and, most important, their communication with each other (Sharma 2016).

Telegram

ISIS, Al-Qaeda and other terrorist groups strongly increased their use of other encrypted social media platforms, most notably Telegram, in order to avoid suspension. One of its biggest advantages is that its content cannot be indexed by search engines. It is an application for sending text and multimedia messages on Android, iOS and Windows devices. The Telegram company is so confident of its security level that it offered twice a reward of $300,000 to the first person who could crack its encryption. Telegram can be used on any device with Internet access including desktops, laptops, androids, iPhones and iPads. Additionally, Telegram also offers several interests, such as coordinating groups of up to 5000 members, message encryption, unlimited media sharing as well as message destruction within a period of time set by the user. Below, we can see how rapidly its users increased since 2013.

Mail2Tor

Mail2Tor is an anonymous email service that allows a user to send messages to anyone. It is a more private service that the user can use through a typical email provider such as Gmail, AOL or Yahoo mail. This service came from the Tor Project and when somebody wants to use it, they firstly have to download the Tor software. As a result, it keeps its users out of network surveillance, threatening the privacy, relationships or other confidential businesses.

Signal

Signal is an encrypted communication application for Android and iOS. It uses the Internet to send one-to-one and group messages. It can include files, voice notes, images and videos and make one-to-one voice and video calls. Signal uses standard cellular mobile numbers as identifiers and uses end-to-end encryption to secure all communications to other Signal users.

The applications include mechanisms by which users can independently verify the identity of their messaging correspondents and the integrity of the data channel. In addition, a desktop client has been released that can link with a Signal mobile client. Many terrorists prefer this tool rather than Telegram, as it is believed to be more secure.

Surespot

Surespot is an encrypted messaging system, being utilised by ISIS and Al-Qaeda. It allows users to have multiple identities on just a single device and is not linked to the users' phone numbers or email accounts, permitting also the sending of voice messages.

Kik

Kik is an instant encryption messaging system app that can be downloaded on any android-operating mobile system through Google Play Store and App Store. Kik preserves users' anonymity, as it allows the users to register without providing a telephone number. It uses a smartphone's data or a Wi-Fi connection to transmit and receive messages, photos, videos, sketches, mobile webpages and other content after users registers a username. It is widely used by ISIS and Al-Qaeda.

Wickr

Wickr is another encrypted messaging app that ISIS has reportedly been using. Wickr protects users' messages and identity through a multilayered peer-to-peer encryption system. Similar to Telegram, its features include also the sharing of encrypted files to other Wickr users with expiration dates. Wickr also allows users to exchange photos, videos and file attachments, and it is available for the iOS, Android, Mac, Windows and Linux operating systems. It is also preferred by ISIS and Al-Qaeda.

WhatsApp

WhatsApp messenger is one of the leading instant mobile messaging applications. It utilises end-to-end encryption and is available for free for all androids, iPhone and BlackBerry users. As it allows for a secure communication to take place, ISIS, Al-Qaeda and other terrorist organisations operatives and its supporters are known to utilise the communication platform to conduct its activities and for direct communication. Apart from that, WhatsApp is also used to spread online propaganda.

Zello

Zello is a push-to-talk app for mobile devices and PCs. It is a Russian-developed, Texas-based encrypted communication app. It is very popular amongst the ISIS jihadist communication. In fact, it was reported that the terrorists' communication for the attack at Stockholm last year took place through Zello. Zello's users can create private or public channels in order to chat easily with others and listen to conversations. There are two versions, the main Zello app that is free and the Zello@Work app that is fee-based.

3.2.7.2 *Internet Memes (JPEGs or GIFs), Steganography and Watermarking*

The use of memes is combined with a cryptographic method known as "steganography". Essentially, it enables an Internet user to hide a message within a message. Digital images encoded as JPEGs or GIFs can in theory be used to carry other data with them using an unsuspicious subject title. High-tech terrorist groups such as ISIS, etc. are using techniques such as steganography (the hiding of messages in images) and watermarking for communicating covertly with each other. In addition, they use Morse codes or DTMF audio files to send confidential codes as well as barcodes or QR Codes to share GPS coordinates or location, maps and automessages. The barcode generally is 12- to 20-digit number and is primarily used for serial numbers, pricing and inventory control of the products worldwide. Barcodes or quick response codes may also be used for communication as well. If any terrorist group wants to communicate via covert communication, they can use this technology as a secure message passing system (Paganini 2016b).

MuslimCrypt

MuslimCrypt is a steganography and encryption tool that was released on January 2018 and performed in order to hide messages inside images. The wide use of MuslimCrypt makes it almost impossible for law enforcement agencies to check every picture on the social media platforms such as Facebook, Twitter, Telegram, etc., plus the actual encryption and steganography algorithms used in MuslimCrypt are still unknown (Fig. 3.9).

3.2.7.3 *Video Games*

Communication through online video games (WoW, for instance) is an increasingly popular way of disguising messages in seemingly unsuspicious interchanges between online "gamers". For instance, PlayStation 4 is even more difficult to keep track of comparing to WhatsApp. PlayStation 4 offers a range of ways to communicate secretly. It supports voice calls between players and online chat messaging. Also, terrorists could communicate without speaking or writing a word by exchanging secret messages within specific games. For instance, terrorists could spell out an attack plan in Super Mario Maker's coins and share it privately with a friend, or two Call of Duty players could write messages to each other on a wall in a disappearing spray of bullets (Bakalar and CBSN 2015).

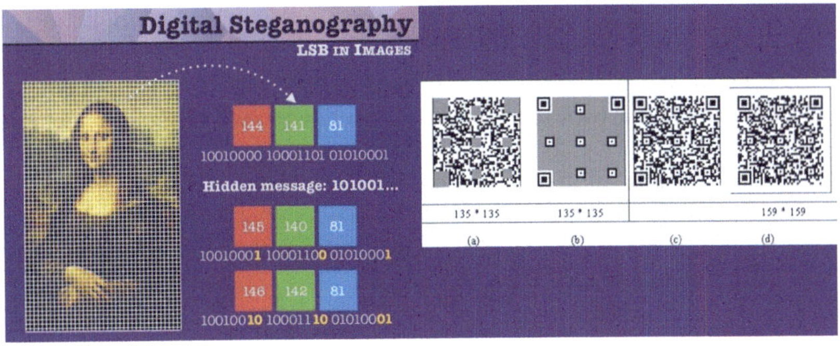

Fig. 3.9 Altered images with hidden content (jgr33nwood 2017; Chen and Wang 2009)

3.2.7.4 Email Dead Dropping

A simple online email account may be used by terrorists for electronic, or virtual, "dead dropping" of communications. It is a technique that has been known for a long time, used to draft emails without sending them, so that different people using the same email credentials can read them without running the risk of interception by third parties.

3.2.7.5 Cybersecurity of Voice Communication

It can be achieved with the use of Linphone, Silent Circle or RedPhone that protect the anonymity and the encrypted protection of voice calls.

3.2.7.6 Terrorist Media

Terrorists use different platforms in their communication and switch between them very often to confuse a potential law enforcement investigation. Al-Qaeda and IS have gone as far as to develop their own tools. More recently, terrorists, like other criminals, are exploiting the opportunities for secure communication provided by smartphone applications and other software, in order to avoid the development and maintenance of their own tools. Although the use of these tools seems to be waning, it is worth to report them as terrorists can rely on them and communicate with each other at any time.

These tools are (Paganini 2014):

- Mujahideen Secrets: the original Mujahideen Secrets (Asrar al-Mujahideen) encryption software launched in 2007, primarily for use with email. Until today there were multiple releases distributed by the Global Islamic Media Front. It is an encrypted email application for Microsoft Windows. It was publicly offered to supporters of Al-Qaeda as a tool to protect the confidentiality of their electronic messages. This software allows users to encrypt and decrypt text messages and files with a range of encryption techniques. This is primarily to ensure that any parties intercepting the messages during transmission, such as via Internet email or cellphone, cannot easily view the message's contents.

- Asrar al-Dardashah: released by GIMF and Al-Qaeda in February 2013, it is an encryption plugin for instant messaging based on the Pidgin platform – which connects to major US-based platforms.
- Tashfeer al-Jawwal: this is a mobile encryption program, again from GIMF and Al-Qaeda, released in September 2013, based on Symbian and Android.
- Amn al-Mujahid: this is an alternative encryption program released in December 2013 from Al-Qaeda's Al-Fajr Technical Committee (FTC). Amn al-Mujahid provides strong encryption that can be easily developed and updated. This program has been provided with a 4096 bit public key, making it the most secure system amongst the other (encryption) algorithms.
- Alemarah: this was an Android-based news app for "terrorist-related actions" that was created by the terrorist group the Taliban and removed from Google's Play Store a few days after it was launched. The purpose of this app was to spread the militant group's message around the world. The main content included videos, recorded from the Taliban in Pashto language.

3.2.7.7 Social Media and Surface Web

Beyond the Dark Web and encrypted communication, the Surface Web is still used by terrorists at a high rate. They use popular social media sites such as Twitter and Facebook in order to spread their messages to the desirable public audience. However, they do not use them only for propaganda spreading but also for the recruitment of jihad fighters, as well as to be embraced by their supporters and justify to them their violent actions. In addition, social media play a very important role to recruit members, gather intelligence and aid the communication amongst terrorists and their supporters during the first phases. Terrorists use mainly the following social media – surface websites – for their communications:

Facebook

Through fake profiles terrorists send numerous friend requests to Facebook users in order to share relative information and message them directly, in order to find new interested members or supporters.

YouTube

Terrorists use YouTube through videos demonstrating the acts (attacking towns, firing weapons) of terrorist organisations such as ISIS, Al-Qaeda, etc.

Twitter

Terrorist follows on Twitter and tweets with propaganda content. Terrorists escort this content with hashtags referring to current events unrelated to terrorism, such as political rallies or sports matches, in languages of regions of interest, including Europe and other Western countries, but also hashtags used by ISIS opponents in the Middle East, to achieve wide dissemination.

Instagram Instagram

Photos or videos are underscored with #hashtags that refer to terrorist organisations or other websites full of propaganda or recruitment preparation. Instagram is also used to post videos and pictures to project the images of terrorists' attacks and brutality.

Tumblr

This online media platform is part microblogging and part social networking. It was founded in 2007 and it hosts 300 million blogs and up to 100 billion posts. Tumblr does not need one to sign up to be able to browse the blogs on the site, unless one wishes to comment or follow blogs. So easily, typing the "right" blog, an immediate communication can be achieved.

Ask.fm

This social media platform has been used sometimes by Daesh to communicate with its members and potential recruits as well as supporters.

Pinterest

Pinterest is also widely used by jihadists to share extremist ideas, articles, images and videos.

Generally, "the crime scene of the 21st century leaves a trail of digital footprints rather than physical ones and the same tools that protect us are being exploited by those on the wrong side of law to cover their tracks" (Harsh 2015).

3.3 Online Terrorist Recruitment

Terrorists do not use the Internet only to communicate, spread propaganda and extremist content (images, video, etc.), but also, it is a way to make relationships with people that are responsive to the massive propaganda and recruit new supporters and fighters. The Internet is a totally effective medium for the recruitment of minors and young people, who make up the biggest part of Internet users. For this reason, terrorists are trying to attract and develop relationships with people, rather than just waiting for them to present themselves (Weinmann 2004). Well-known terrorist organisations such as ISIS are so experienced to the use of social media for propaganda and recruitment, and their online activity can easily be traced to social media platforms. For instance (Harsh 2015), the spreading of radicalisation on social media is causing increasing concern with a reported 90,000 Twitter accounts being controlled by ISIS to target and recruit young people and isolated teens into a war where hashtags are becoming the new weapons. Nowadays, all these are being monitored closely, forcing in this way terrorist organisations to search for an alternative safe online action. It has come up that the Dark Web is the perfect solution as it is completely anonymous.

3.3.1 *Surface Web*

The Surface Web is a valuable tool for terrorists in order to spread their propaganda and recruit new members affiliated to their goals. These goals so far could be easily achieved with the maximum results by using social media platforms such as Twitter and Facebook (Paganini 2016a).

3.3.2 *Social Media (Fig. 3.10)*

Terrorists are using widely in the last years the social media to target thousands of young kids and early adults (Taylor 2016). Indeed, according to researches, ISIS and other terrorist organisations are using platforms such as Facebook, YouTube and Twitter to reach their target groups (usually isolated young people) with attention attracting posts. For example, "they may post a photo where a terrorist has a kitten in one hand and an AK-47 in the other hand", said John Carlin, assistant attorney general for national security at the Department of Justice.

Terrorists use social media platforms to share their propaganda and to create a sense of common purpose for supporters, especially when attempting to recruit young people. New interested people might become online friends with several different terrorist supporters. With the frequent interaction, these young isolated and marginalised people start to watch the terrorism supporting posts, and after a lot of communication, they feel like that they are important to someone, their opinion counts and they are a

Fig. 3.10 Attraction of potential recruits by posting such photos on social media (Woods 2016)

Fig. 3.11 Social media use for recruiting (Verton 2014)

part of a community that has a wide, upper, important purpose. We could say that this is the step of radicalisation and the step of a potential recruitment (Fig. 3.11).

3.3.3 *Internet Websites and Chat Rooms*

Beyond social media, terrorist organisations also build their own websites, chat rooms, forums and online magazines. With the expanding use of the Internet, the number of terrorist home-built websites increased exponentially. More specifically, 20 years ago, there were 12 terrorist websites active. Nowadays, this number has outreached 7000. In addition, terrorists and their supporters administrate many profiles on different social media websites in order to spread the propaganda, advertise the group's activities and attract new supporters. Combined with the increasing trend of the Internet and the rising of technology, recruitment has become much easier and quicker than ever. Terrorists do not wait for the interested people to contact them. They seek information about the people that browse their websites, and if they believe that these people have the necessary profile relating to the group's purposes or are able to carry out any group's work, they are immediately contacted. Furthermore, lots of these websites concern "the training" of the recruited members. Any potential recruit who will be given access to these restricted sites has the ability to find plenty of material and be instructed by terrorists on how to build explosives, how to execute specific terrorist attacks, how to obtain firearms and how to join a terrorist organisation. Also, potential recruits receive numerous religious messages and propaganda, provided with training manuals on how to be a terrorist.

Finally, there are also websites created by terrorists that are addressed to minors (Dodds 2011). In this case, propaganda is disseminated with the form of cartoons which promote acts of violence and terrorism. Terrorist organisations such as Hamas and Al-Qaeda have attempted to recruit children using Disney-style animated videos (Kaczynski 2013). A few years ago, Al-Qaeda released a short film featuring cartoons – young boys dressed in battle suits, participating in terrorist attacks.

3.3.4 Video Games

Like home-built websites, many terrorist organisations have designed and developed online video games, popular music videos and computer games for young people recruitment and training purposes. These video games promote mainly the use of violence against Western countries and their political people, offered in multiple languages, in order to appeal to a broad audience and especially teens and young adults, who comprise maybe the highest proportion of users. For instance, Hezbollah (CNN 2007) released the games Special Force and Special Force 2, which depict themselves fighting the Israeli military. The Global Islamist Media Front, in association with Al-Qaeda, released the Quest for Bush game online. The game, aimed at children, sets the goal of killing the president of the USA.

Generally, the Surface Web provides terrorist organisations and sympathisers with a global pool of potential recruits, mainly young members of vulnerable and marginalised society groups, filled with sentiments of injustice, exclusion or humiliation. Moreover, the use of technological barriers to enter recruitment platforms (through invitation, credentials, etc.) also increases the complexity of tracking terrorism-related activity by intelligence and law enforcement personnel.

3.3.5 *Covert Recruitment and Training*

As mentioned previously, in the beginning, terrorists used the Surface Web in order to make their presence perceivable and to disseminate the propaganda, hoping of finding new supporters and potential new recruiters for the implementation of their purposes and plans. So, it is easy to understand that a large amount of things happen publicly on Surface Web, via social media, online magazines, chat groups, etc. When the time of communication with the potential members is about to come, terrorists increase their interaction, and when the discussions become more pointed, the communication turns to be conducted through encrypted apps, for the secrecy and to avoid the disclosure of its content.

The Dark Web is more and more expansive, and it is already difficult for law enforcement agencies to make constant investigations on it and its well-hidden unindexed by search engines content. Therefore, big terrorist organisations such as ISIS, Al-Qaeda, etc. take advantage of it and have started to use more and more extensively the Dark Web for their purposes, such as propaganda, recruitment after they have approached the new nominated members or supporters and training. Websites within the Dark Web allow terrorists to recruit young men, hire hitmen, purchase stolen credit cards and fake IDs in order to pose their plans easily and remain undetected by law enforcement. Beyond recruitment Dark Web is also an ideal place for terrorists' attack plans. Therefore, it is a great challenge as all the tactics and ways of a terrorist organisation planning, implementation and execution can be found in Dark Web. Terrorists can now "go dark" by using also strong encryption (Crime 2012; Allen 2015). This can be done very easily with already famous end-to-end encryption apps such as WhatsApp (Weise 2017) and Telegram. Both of them are widely used so far for communication and coordination during terrorist attacks. The secrecy that they provide is totally comfortable for the constant touch between terrorists and the recruitment of new members. After the completed recruitment of the new member, they are guided by the terrorists to join websites within the Dark Web, by using the specialised browsers. In the Dark Web, there are a lot of websites, most of them encrypted and available only to those who are given the instructions about how to find them and how to access them. In this stage, the Dark Web plays a very important role for the new terrorist group recruiters. The overall communication strategy of a terrorist group is contacted and depends on these protected websites. In addition, in these websites, new recruits are able to find "training" material for constructing bombs, for attack plans and other useful things that are part of explicit terrorist tutorials. One of the most valuable training tools is to learn how to run all the traffic on their Android mobile phones through Dark Web. In that way, it is assured that all their Internet and voice traffic is sent through encrypted channels, remaining undetected and unencrypted by law enforcement.

This is the main advantage that the Dark Web provides to terrorists: an anonymous network readily available yet generally inaccessible. It would be extremely difficult for terrorists to stay on the Surface Web without being undetected or their websites, posts and relative material being shut down by law enforcement agencies. That is why terrorist organisations turned to the Dark Web; they liked the idea of remaining on schedule concerning their purposes without worrying for their identification and the deactivation of their websites, while at the same time they enjoy an easier and totally secret communication. Any terrorist website on the Surface Web would be instantly shut down, and its admin would be arrested. On the Dark Web, decentralised and anonymous networks aid in evading arrest and closure of these terrorist sites.

3.3.6 *Example of Lone Wolf Terrorists Online Recruitment*

The recruitment of lone wolf terrorists relies on online platforms and requires a gradual transition through numerous phases as presented below (Weinmann 2014):

- The net: all online platforms are used at this stage by terrorist organisations, such as official websites, Facebook, YouTube, Twitter and other social media. At this point, recruiters view the whole population as primed for recruitment and expose it to an online message, video, taped lecture or word document. The approach occurs at a common way in order to filter the positive responses from the negatives.
- The funnel: at this stage a potential recruitment has occurred, yet needs a significant transformation in identity and motivation, which is succeeded through a virtual social bonding, based on the target's negative feelings, such as social frustration, solitude and personal pessimism. Recruiters seem very friendly to the target and that they have a real interest in them, while at the same moment, they expose

them to religious, political or ideological material achieving their initial radicalisation.
- The infection: this stage includes advanced radicalisation by continuous exposure to online radical material and virtual online guidance which lead to the preparation and recruitment of the target.
- The activation: this final step includes practical instructions and training through online material to make and use explosives and weapons as well as direction regarding the details of a terrorist attack.

3.4 Terrorism Funding

Nowadays terrorists, in order to prepare and implement their attacks, need to have at their disposal a lot of resources and equipment. Travel costs, car and safe house rental, weapons and explosives require a lot of money. Under any circumstances, even a big terrorist organisation such as ISIS cannot handle these expenses or finance all the attacks that take place in EU countries, for instance. It has been proved through researches that all the recent EU terrorist attacks have been funded by legal and illegal sources. Terrorists deploy various techniques to raise and use funds and move them through different tools, financing their acts of terrorism. The methods used on the Internet in order to raise and collect funds and resources by terrorists can be classified in the following four general categories (Conway 2005):

- Direct solicitation: it includes the use of websites, chat, chat groups, massive emails and other means of communication in order to request donations from supporters.
- E-commerce: many websites are used as online stores (selling books, audio, items, etc.) to hide their real activity from third-party people.
- Exploitation of online payment tools: online payment facilities offered through dedicated websites with credit cards or payment facilities such as PayPal or Skype. Online payment can also take place with fraudulent mean (identity or credit card theft, fraud, etc.).
- Through charitable organisations: charities provided to seemingly legal organisations (e.g. humanitarian, philanthropic, etc.) can be used for cover in order to perform illegal purposes such as promotion and funding of terrorist acts.

In addition, it came out that terrorist organisations can also be financed by several types of crime such as drugs, thefts, robberies, counterfeiting and burglaries. Also, some terrorist groups are familiar with computer science, and many of their members or supporters through cyber techniques finally distract some amounts from financial transactions. In addition, Internet demographics allow terrorists to identify users with sympathy to their activities. The terrorists contact these users by email or social media, and usually they find them willing to donate to the terrorist organisation.

3.4.1 *Funding and Buying in the Dark Web*

Nowadays, terrorists tend to use the Dark Web more and more for fund-raising, money transfers and illegal purchase of explosives and weapons, using virtual currencies such as Bitcoin. Bitcoin was introduced in 2008, as a type of an alternative currency, but its ability of avoiding detection inspired many terrorist organisations, such as ISIS and Al-Qaeda, to adopt it as a common way of transacting and funding. So, it is strongly believed that the biggest part of donation and terrorism funding is achieved by transferring Bitcoins to terrorist organisations' Bitcoin wallets. In fact, there are plenty of pages that can be found on the Dark Web which usually invite jihad supporters to donate to terrorist organisations through transactions to a particular Bitcoin address. Many of them contain files-documents as guidelines so a user is able to complete a Bitcoin transaction using a Dark Wallet avoiding detection from authorities. All this aims to the setup of a globally untraceable Bitcoin donation system. For instance, one Dark Web page is called "Fund the Islamic Struggle without Leaving a Trace", and it invites donators to complete transactions in a particular Bitcoin address. A PDF document is posted online as a guide for using the Dark Web for secretive financial transactions as well. This document refers to the Dark Web black markets and explains the procedure that someone can follow to buy illegal items such as weapons using Bitcoin and the Dark Wallet application.

It is strongly believed that the Dark Web is used by terrorists' organisations in order to buy weapons and explosives for their massive attacks, such as the attack in Paris. Terrorists do not use the Dark Web only for buying items or to receive donations. Some reports correlated that terrorist organisations and especially ISIS (Berton 2015) use the Dark Web for selling too, such as human organs (probably from their hostages) or stolen antiquities and other precious items and goods, such as oil (ISIS tactics) in order to raise money and spend it on their activities. In addition, terrorists try to raise money through extortion and demanding ransom for their captives. All these funds coming from the Dark Web permit both terrorists and their supporters to remain anonymous and undetected (Fischer 2015). For example, there is a Dark Web platform named "EuroGuns", in which there are photos of various weapons posted for sale; the value of most is attractive and rational (e.g. AK-47 is sold for $550). In addition, books such as a terrorist's handbook and explosive guides can be purchased through dark markets. It is possible for terrorists to "visit" a dark market on the Dark Web where they can buy fake documents and passports, driver's licenses, ID cards, stamps and other products for use in the UK, the USA and other countries where they desire to make terrorist attacks.

Terrorists do not use the Dark Web only to communicate secretly or receive money through cryptocurrencies. They "visit" the illicit marketplaces on the Dark Web in order to buy and sell all the necessary equipment for their operational needs without being detected from law enforcement or governmental agencies. Some of this equipment includes fake passports (for an easy cross border, rent of houses and cars and other commercial goods). In some cases it is believed that terrorists have employed cybercriminal groups (Maxey 2017) to provide them with their technical expertise in order to launder cryptocurrencies. Furthermore they help them conduct big operations, such as ransomware attacks, which is a very profitable business. On the other side, cybercriminal groups may not even know they help terrorists as the cryptocurrencies provide the desired anonymity. For instance, the cryptocurrency Monero does not indicate the sending and receiving accounts or the amount transferred on public ledgers, so it could be a very convenient currency for terrorist groups.

For all the above activities, an ISIS supporter created a guide (Bertrand 2015) where it explained how anyone could fund terrorists using a Dark Wallet, which is a Dark Web app that protects and anonymises the crypto-transactions. The same thing happened with another ISIS supporter calling himself Amreeki, who posted a tweet referring to a PDF file in WordPress indicating the importance of funding terrorists through Bitcoin transactions, as the other ways are traceable and easy to be interrupted by governments around the world (Fig. 3.12).

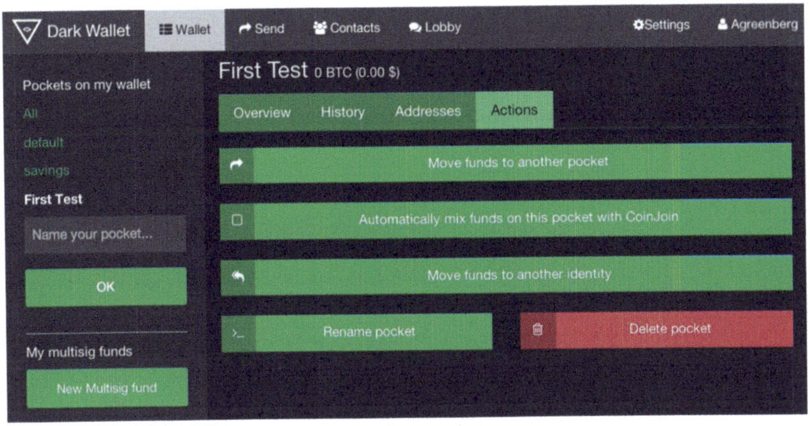

Fig. 3.12 Screenshot of a Dark Wallet (Bertrand 2015)

3.4.2 Moving Terrorist Funds and Donating to Terrorist Organisations

Big terrorist organisations, in order to secure and protect their funding, were forced to start using cryptocurrencies such as Bitcoin as worldwide governments managed to stop traditional banking methods being used to fund terrorist organisations. Indeed, the last years, dark markets such as Silk Road, etc. were used by terrorists to purchase weapons. In addition, through these dark sites, any user could donate to terrorist organisations. While terrorist groups such as the real-IRA, Al-Qaeda and JI had used the Internet to collect funds through online donations and charity organisations, ISIS evolved online funding at another level, including retailing of prisoners, donations through social media platforms and the use of Bitcoin (Sewerd 2015). Moreover, ISIS took under control many banks in the territories that it captured, and it is estimated that in their hands fell over 400 million dollars. Apart from all that, ISIS runs their own black market for selling oil and weapons or even supplies other organisations with similar ideology (Alkhouri 2015).

As far as online funding is concerned, ISIS conducts online fundraising, both on the Surface and the Dark Web. Especially in the Dark Web, many ISIS website Bitcoin accounts are provided to raise funds from their supporters with appealing messages that call Muslims from all over the world to fund them via Bitcoin. Also, in some websites, there are phone numbers posted, urging supporters to contact them directly for donations. The same happens with fundraising campaigns that are organised via sending messages to donors on Telegram. It is not specified how the donation must be done, but donors are urged to communicate directly to receive further instructions.

3.4.3 *Cyberterrorism*

Due to the fact that terrorists have limited sources of fund, crimes such as cyberattacks are more and more attractive as they require much smaller number of attackers and certainly smaller amounts of money in order to perform them. In addition, the terrorists who have profound knowledge of computers know how to cover their "traces" when performing the cyberattacks and remain unknown and undetected both geographically and technologically. The biggest advantage of cyberattacks is that cyberterrorists can perform them from anywhere preserving at the same time their anonymity. It is strongly believed that cyberterrorism has the most effective result when it is combined with physical terrorism. Possible targets of cyberterrorism could be government computers and financial networks to cause chaos to the public. Vital data and resources, such as classified information, data deletion, website damaging and viruses inserting, are just a few ways of how terrorists can enter into a secured system and cause damage. Cyberattacks are used so that terrorists can raise the financial funds and distribute their propaganda.

Finally, terrorist groups can, at times, resort to common crime to generate funds which are then used to cover the costs associated with the planning and execution of attacks, such as recruitment, procurement and travel. Alternatively, they may seek contact with common criminals or organised crime groups to access greater financial resources, weapons, transport means, specialist skills or a larger pool of potential recruits.

3.5 Concluding Remarks

The use of the Dark Web by a plethora of terrorists is a very challenging situation for our era and the worldwide peace. That is because terrorists have found the way to slip from law enforcement agency's persistent surveillance and implement their plans in an easier way than before. This gives a big advantage to terrorists as they cannot be traced and their online activity cannot be monitored and removed. As a result, more and more people are influenced by terrorists and support their goal. The encrypted communication that Dark Web and the rest of the end-to-end encryption apps such as Telegram and WhatsApp provide poses a severe concern for governments and the law enforcement agencies, as in most cases they are unable to spring on terrorists after detecting their communications and position. This is a big trouble as we will not be able to know what terrorists plan, before it happens.

The Dark Web assists the covert activity of terrorists and their supporters, concerning their attack plans and the funding of all their activities. The fact is that terrorists are unbothered to accomplish their financial trades and transactions and get more power, by receiving money from supporters who are buying weapons and equipment.

The crucial question is: *should governments be able to access the content in messages exchanged through these encrypted apps*? The previous experience on this case says that this could generate greater danger. The creation of a back door would probably lead to data breaches, sacrificing our privacy. Moreover, if this would happen, terrorists would find an alternative way to communicate with each other and implement their tactical and financial plans.

So, with the restrictions that the Dark Web and encrypted apps impose, we should try to find an effective way to restrict the terrorists' online action. It is a great challenge though; big European organisations such as Europol, Interpol and law enforcement agencies by coordinated cooperation through big operations and cyber patrolling manage to remove profiles, pages, websites and chat groups with propaganda and extremist content, trying to avoid the dissemination of terrorist ideology through the social media that most people use daily. Concerning the Dark Web, the landscape is obscure, but the organised and consistent operations could have better and better results to this daily combat against terrorism and extremist violence.

References

L. Alkhouri, *Bankrolling Terror: The Funding & Financing of ISIS* (2015)

E. Allen, *ISIS Uses Deep Web to Raise Money, Recruit Fighters* (2015)

J. Bakalar, CBSN, *Terrorists Use Video Games to Communicate Undetected.* [Sound Recording] (CBS News) (2015)

BBARCESAJ, [Online] (2013), Available at: http://www.tootips.com/2013/07/bitmessage-decentralised-p2p-network.html. Accessed 2018

BBC News, Paris attacks: What happened on the night. *BBC News* (2015)
K. Berrada, M. Boudier, Can ISIS's cyber-strategy really be thwarted?. J. Strateg. Threat. Intell (2017)
B. Berton, [Online] (2015), Available at: https://www.iss.europa.eu/content/dark-side-web-isil%E2%80%99s-one-stop-shop
N. Bertrand, *ISIS is Taking Full Advantage of the Darkest Corners of the Internet* (2015)
Chat Onion, *OnionApps/Chat.onion.* [Online] (2016), Available at: https://github.com/onionApps/Chat.onion. Accessed 2018
W.-Y. Chen, J.-W. Wang, Nested image steganography scheme using QR-barcode technique. Opt. Eng **48**(5)4-6, (2009)
H. Chen et al., *Terrorism Informatics – Knowledge Management and Data Mining for Homeland Security* (Springer, s.l., 2008)
M. Chertoff, T. Simon, *The Impact of the Dark Web on Internet Governance and Cyber Security*, vol 6 (2015)
CNN *Hezbollah Video Game: War with Israel* (2007)
M. Conway, *Terrorist 'Use' of the Internet and Fighting Back* (Department of Political Science, College Green, Trinity College, Dublin, 2005)
Crime, U. N. O. o. D. a., *The Use of the Internet for Terrorist Purposes* (United Nations, New York, 2012)
P. Dodds, Al-Qaeda plans cartoon recruiting film. *The Sydney Morning Herald* (2011)
Europol, *Changes in Modus Operandi of Islamic State (IS) Revisited* (Europol, The Hague, 2016)
Europol, *EU Terrorism Situation and Trend Report (TE-SAT)* (Europol, The Hague, 2017)
Europol EC3, *Internet Organised Crime Threat Assessment* (Europol, The Hague, 2016)
Europol EC3, *Internet Organised Crime Threat Assessment* (Europol, The Hague, 2017)
K. Finklea, *Dark Web* (Congressional Research Service, s.l., 2017)
J. Fischer, *The Bitcoin-ISIS Connection.* [Online] (2015), Available at: https://www.atmmarketplace.com/articles/the-bitcoinisis-connection-2/
F. Gardner, How do terrorists communicate? *BBC News* (2013)
Gate Stone Institute, [Online] (2017), Available at: https://www.gatestoneinstitute.org/pics/3198.jpg. Accessed 2018
Guardian Project, *Orbot: Tor for Android.* [Online] (2016), Available at: https://guardianproject.info/apps/orbot/. Accessed 2018
A. Harsh, [Online] (2015), Available at: https://www.linkedin.com/pulse/how-terrorists-communicate-dark-web-anurag-harsh

INSITE Blog on Terrorism & Extremism, [Online] (2015), Available at: news.siteintelgroup.com/blog/images/easyblog_images/953/November2015. Accessed 2018

IntelCenter, *Islamic State's 43 Global Affiliates – Interactive World Map*. [Online] (2016), Available at: https://intelcenter.com/maps/is-affiliates-map.html. Accessed 2018

Israels ACG, *Twitter*. [Online] (2018), Available at: twitter.com/IsraelsACG/status/667937641225146368. Accessed 2018

jgr33nwood, *Steganography and Cybercriminals: Hidden in Plain Sight*. [Online] (2017), Available at: https://steemit.com/bitcoin/@jgr33nwood/steganography-and-cybercriminals-hidden-in-plain-sight. Accessed 2019

A. Kaczynski, 8 Ways terrorists use the internet for recruitment. *BuzzFeed News* (2013)

Krypt3ia, *The First Official Da'esh DARKNET Bulletin Board Has Arrived*. [Online] (2015), Available at: https://krypt3ia.wordpress.com/2015/11/15/the-first-official-daesh-darknet-bulletin-board-has-arrived/. Accessed 2018

L. Maxey, Terror finance in the age of bitcoin. *The Cipher Brief* (2017)

MEMRI CJ LAB, [Online] (n.d.), Available at: http://cjlab.memri.org/lab-projects/monitoring-jihadi-and-hacktivist-activity/pro-isis-telegram-account-posts-dark-web-advertisement-for-hiring-albanian-assassins-and-hackers-you-dont-have-to-pay-till-the-job-is-done/

Opinion, [Online] (2017), Available at: http://opinion.al/author/trim/page/24/. Accessed 2018

P. Paganini, Al-Qaeda is developing new Encryption tools in response to NSA surveillance. *CDM – Cyber Defense Magazine* (2014)

P. Paganini, [Online] (2016a), Available at: https://securityaffairs.co/wordpress/45755/terrorism/dark-web.html

P. Paganini, [Online] (2016b), Available at: http://securityaffairs.co/wordpress/47243/terrorism/covert-communications-terrorists.html

Quilliam Foundation, *Documenting the Virtual 'Caliphate'* (Quilliam Foundation, s.l., 2015)

D. Reisinger, The many ways terrorists communicate online. *Fortune* (2016)

C. Sewerd., https://www.liquidvpn.com. [Online] (2015), Available at: https://www.liquidvpn.com/isis-use-tor-and-bitcoin-to-fund-terrorism/

M. Sharma, *Dark Web and Encrypted Apps: ISIS Communicates in the Black* (Institute for Defence Studies and Analyses, 2016)

Site Staff, *IS Shifts Propaganda Archive to the Dark Web* (SITE Intelligence Group, 2015)

Stratfor, *As-Sahab: Al Qaeda's Nebulous Media Branch*. [Online] (2016), Available at: https://www.stratfor.com/analysis/sahab-al-qaedas-nebulous-media-branch. Accessed 16 Dec 2017

H. Taylor, [Online] (2016), Available at: https://www.cnbc.com/2016/10/05/most-young-terrorist-recruitment-is-linked-to-social-media-said-doj-official.html

D. Verton, *Are Social Media Companies Doing Enough to Stop Terrorist Recruitment?* [Online] (2014), Available at: https://www.fedscoop.com/social-media-companies-enough-stop-terrorist-recruitment/. Accessed 2018

G. Weinmann, *How Modern Terrorism Uses the Internet* (United States Institute of Peace, Washington, 2004)

G. Weinmann, [Online] (2014), Available at: https://medium.com/its-a-medium-world/virtual-packs-of-lone-wolves-17b12f8c455a

E. Weise, Terrorists use the Dark Web to hide. *USA TODAY* (2017)

Wired, *Tor Just Launched the Easiest App Yet for Anonymous, Encrypted IM.* [Online] (2015), Available at: https://www.wired.com/2015/10/tor-just-launched-the-easiest-app-yet-for-anonymous-encrypted-im/#slide-2. Accessed 2018

J. Woods, *Why teenage girls can't see beyond Isil's 'cute kittens and Nutella' ploy.* [Online] (2016), Available at: https://www.telegraph.co.uk/women/life/why-teenage-girls-cant-see-beyond-isils-cute-kittens-and-nutella/. Accessed 2018

CHAPTER 4

Dark Web Markets

Emmanouil Kermitsis, Dimitrios Kavallieros, Dimitrios Myttas, Euthimios Lissaris, and Georgios Giataganas

4.1 Introduction

Dark Web markets, darknet markets, dark markets, black markets and crypto-markets are some of the new terms that have been introduced in the recent years referring to all the online illicit marketplaces that have been developed and operate in the Dark Web environment. These relatively new online marketspaces are very similar to the existing marketplaces in the Surface Web like Amazon, eBay, etc. providing though illegal goods, trading services and transaction facilities to their customers under

E. Kermitsis • D. Myttas • E. Lissaris • G. Giataganas
Center for Security Studies-KEMEA, Athens, Greece
e-mail: e.kermitsis@kemea-research.gr; d.myttas@kemea-research.gr; e.lissaris@kemea-research.gr; g.giataganas@kemea-research.gr

D. Kavallieros (✉)
Center for Security Studies-KEMEA, Athens, Greece

University of Peloponnese-Department of Informatics and Telecommunications, Tripoli, Greece
e-mail: d.kavallieros@kemea-research.gr

B. Akhgar et al. (eds.), *Dark Web Investigation*,
Security Informatics and Law Enforcement,
https://doi.org/10.1007/978-3-030-55343-2_4

advanced encryption protecting the anonymity of their users and making consequently the whole market totally anonymous.

The underground economy is the emerging and increasing trend in the dark market transactions. Among all the Internet and mobile technologies that are continuously evolving, enabling the criminals to operate in the darknet market world, digital currencies known as cryptocurrencies are some more of such digital tools which are currently transforming our criminal underworld facilitating the payment of transactions between sellers and buyers for their illegal trading activities and posing threats for money laundering and terrorism financing. Before transactions are initiated, money should first be converted to cryptocurrencies.

4.2 Characteristics and Features

When a product or substance changes its status from legal to illegal, there is usually also a sales shift from Surface Web to the darknet marketplaces. VPN service, Tor or I2P browser, PGP encryption/decryption and authentication are some of the main features of these sites for safe access, communication and information sharing. Bitcoin, Monero or other cryptocurrencies are used for safe trading and payments. Special focus is also given by the market developers to the design with a good-looking display and a user-friendly interface of the market.

Most of the dark markets require buyers and vendors to register by filling registration forms with their data. Since anonymity is the primary concern, it is recommended not to fill detailed contact and other information which will reveal the real identities.

Vendors' trust is another significant characteristic for the market's overall trust. Vendors' trust levels, among other factors, depend very much on their activity in the site; the more active a vendor is the higher level of his trust. Many marketplaces provide the chance to buyers to evaluate the vendors with their feedback using star rating or other similar reputation systems; the more neutral and negative the feedback the lower level of trust is developed for the market. For markets which are focusing on vendor's reputation or are relatively new, they present vendor ratings also from other darknet markets.

One phenomenon that has been observed in the darknet markets is the sudden exit or shifts of some markets from the Dark Web. This usually takes place when after a sufficiently large amount of money in cryptocurrencies are shifted to these marketplaces they close the site in order to block outgoing transfers and make off with the money they have locked into it. A suspect of such type of an exit scam was the case of the Nucleus Market which all of a sudden disappeared on April 13, 2016. Many of such sites like the Sheep Marketplace and TheRealDeal market became also suddenly unreachable in 2016 when their owners and admins vanished after absconding millions of dollars' worth of the marketplaces (Brown 2016).

Another observed phenomenon is when a popular darknet market is closed down, all its users are moving to other prominent dark markets increasing in a very short time exponentially their members and activities. When Silk Road 2.0 was taken down in 2015, Agora and Evolution marketplaces increased their memberships by 20%. We had a similar case after the takedown of AlphaBay when many of its users moved to Hansa market. Many of these users have been identified by the Dutch authorities who in the meantime had taken over the control of the Hansa market (Staalduinen 2017) (see Chap. 11).

4.2.1 Goods and Services

According to the content, type of goods and services on the Dark Web sites, the following six major types of dark markets are identified among other Dark Web categories:

- Arms – with trading of firearms, weapons and other related goods illegally sold by firearm traffickers to private individuals, organised crime groups (OCG) and terrorists.
- Drugs – with manufacturing or trading of a large variety of illegal drugs either pharmaceutical or otherwise including illegal prescription medicines. Offers on drugs vary from heroin, cocaine, marijuana, methamphetamines and various forms of acid to markets such as those trading anabolic steroids and Viagra-made medication.
- Jewellery and gold – mainly gold and silver products.
- Fraud and counterfeits – the document fraud, with the online trading of fraudulent, forged, stolen and counterfeited documents and cards, such as fake passports or identification cards and cloned and stolen credit cards or accounts, is emerging and one of the fastest-growing markets, in all types of criminal activities including terrorism. 'Card shops', for example, are one of the specialty markets in the Dark Web.
- Guides and tutorials – for darknet market users on fraud, hacking, drugs and many security-related and anonymity issues.

- Digital goods and services – such as software, source code, botnets and malware, e-books, game keys, hosting, VPN and security services, currency exchanges and financial services.

Most of the markets include one or more of the above categories of goods and services in their listings, whereas those considered as descent dark markets usually have also a list of restricted or prohibited items so that buyers and vendors can know that these items cannot be traded in their marketplaces and in case of any violation of such prohibitions a permanent ban can take place.

According to the Serious and Organised Crime Threat Assessment (Europol 2017b), it is estimated that the top 1% most successful vendors are responsible for 51.5% of all transactions on dark markets. In one study 57% of active sites could be classified as an illicit-related form of activity. Early researches showed also that 80% of the trade consists of drug transactions and the other 20% covers arms, fraud and counterfeits and other digital products and services.

4.2.2 *Multisig or Trusted Markets*

Some markets are considered as multisig markets because they apply transactions with multiple signatures. In contrast with the single-signature transactions in the Bitcoin or other cryptocurrency networks where payments require only one signature, the multi-signature (multisig) transactions require the signature of multiple parts before funds are transferred and therefore require more than one key to authorise such transactions making the payment process safer for all actors. In multisig markets there are three keys generated, which either all or two of them are required to release the payment. One key is with the vendor, the other is with the buyer, and the third is with the market. This category of darknet markets that utilise multisig transactions is also known as trusted markets since they are usually those that have good reputation – with no security issues and no scamming or other technical issues. Wall Street Market, Cannabis Growers and Merchants Cooperative (CGMC) and Point/T.chka are some examples of active multisig markets (DeepDotWeb 2019d).

4.2.3 *Escrow Markets*

Other markets are using regular escrow, where the money required to buy the item(s) is removed from the buyer's wallet and placed in a market 'escrow' account to ensure the required cash will be there when someone takes the purchase order. Dream Market, Berlusconi Market and The Majestic Garden Market are some examples of active escrow markets (DeepDotWeb 2019a).

4.2.4 *'Finish Early' Markets*

Payments in this method are sent directly to the vendor. This method is used with trusted vendors or those with whom the transactions were successful in the past. For first time purchases though, this method is not so safe since there is no prior payment experience with vendors and payments cannot be claimed back in case the purchased items are not received.

4.2.5 *Invite/Referral Markets*

This category is for markets that require an invite code or a referral link in order for the users to register and start their trading and transaction activities. These types of markets usually have an affiliate program giving commissions to users who have referred new users to the site through their respective referral links. There are market cases where different vendors apply different payment methods within a marketplace or payments are arranged between buyers and sellers after they come in contact with each other without the market's intervention and involvement to their transactions. Wall Street Market, Dream Market and Point/T.chka are some examples of active invite markets (DeepDotWeb 2019c).

As we notice, one market place might be included in more than one of the above categories according to the type of registrations and transaction payments they offer.

4.2.6 *Charges, Transaction and Payments*

There are three the main actors who get involved in the transactions of a dark market: the sellers, the buyers and the market itself as the intermediate enabler and service provider bringing buyers and sellers together in one space providing and ensuring anonymity, security and trust among its users.

As soon as a buyer selects a market for a purchase, the next most important point is the selection of the vendor in the market. It is recommended to select those vendors that are active on the site, don't receive a lot of complaints, offer good sales terms, have been operating on the site for a long time and support multisig payments.

Various types of charges and combinations of them are applied to vendors by the markets:

- One-off fixed registration fee when a vendor enters a market.
- Fee which serves as a bond that is returned after a vendor closes the account or once he proves that he is a seller of good standing.
- Rental fees providing vendors the service to create their own stores within the marketplace. Rental fees vary depending on the size of the store.
- Commission percentages on sales made through the market sites.

4.3 Cryptocurrencies

A cryptocurrency, as its name implies, is a peer-to-peer digital or virtual currency that is exchanged by using certain principles of cryptography. It can be used as a normal fiat currency such as the US$ or currency in the country that a person stays but with the big difference as it is not regulated at all by any bank. More than 100,000 vendors and companies currently accept cryptocurrencies as a valid form of payment, and as their value increase, that number seems likely to grow.

As one of the main concerns of cryptocurrencies is their use to facilitate criminal trades and services, they are introducing new types of economic crime organisations hard to be investigated with the use of traditional investigation methods by the police and the financial/tax or other authorities. For this reason new techniques and technologies have been developed to help police identify flows of transactions and trace the criminal use of cryptocurrencies. Blockchain, primarily developed as accounting method for Bitcoin, is one of these technologies used for a digitised, decentralised, public ledger of all cryptocurrency transactions allowing market participants to keep track and verify digital currency transactions. In fact, each completed transaction is publicly recorded into a list of records called blocks which are appended eventually into the blockchain, where they are verified and relayed by other cryptocurrency users and by the entire community using the blockchain instead of a single centralised authority.

Table 4.1 Differences between fiat present currencies and cryptocurrencies

Fiat present currencies	*Cryptocurrencies*
Issued by government	Created by computers
Unlimited supply and can be produced more by the government when necessary	Limited supply and has a set of maximum
Physical way of exchange money	A digital way of money exchange
Form of coin or paper	Presented by private and public piece of code
Centralised and controlled by the law and banks	Decentralised and not controlled by any single entity or government
Its value is determined by the market and regulation	Its value is determined by the supply and demand
Currencies are used within specific geographical borders	Currencies with no borders

4.4 The Differences Between Cryptocurrencies and Fiat Present Currencies

The main differences between the cryptocurrencies and the fiat present normal currencies are given in Table 4.1.

4.4.1 The Differences Between Cryptocurrencies and Digital Currencies

While a digital currency is a currency that represents any fiat currency that is used to transfer between banks and although many of the properties of cryptocurrencies listed in the above table are already due to the fact that cryptocurrency has been defined as a digital currency, main differences between digital and cryptocurrencies are additionally summarised in Table 4.2.

4.4.2 How to Obtain Cryptocurrencies

There are two main ways of obtaining cryptocurrencies: by mining or trading/exchanging.

4.4.2.1 Mining

Unlike fiat currencies that are printed from central governments and banks, cryptocurrencies are created by cryptocurrency miners who are members of the general public. Mining is the term used for creating

Table 4.2 Differences between digital currencies and cryptocurrencies

Digital currencies	*Cryptocurrencies*
Onymous and traceable	Anonymous and encrypted
They are regulated by governments and central banks of a country	They are decentralised currencies
Discrepancy of exchange rates which change on a daily basis	Exchange rates are regulated instantaneously by supply and demand
Limitations on transfer amounts	No limits in transfer funds
High fees charged by banks for fund transfers	Lower fees for fund transfers
Long fund transfer time	Shorter fund transfer time
Total control is in banks' hands	Each person deposits the amounts in his digital wallet. He/she is the banker and has total control and use of his/her funds
No limitations in the creation and circulation of digital currencies	Cryptocurrencies have a limited amount that will be created, and it is predicted that their values will increase for the foreseeable future
Widely spread in the society	Low level of acceptance by the society, related to grey and dark transactions by the public. Not accessible by anybody, specialised knowledge required

cryptocurrencies as a reward to the miners, for proof of work (PoW) involving cryptographic hashes in order for currencies to be verified and prevent them from being abused. The miners compete with each other to solve the cryptologic puzzle (the hash algorithm), which is different for each currency (e.g. for Bitcoin it is the SHA-256 hash algorithm). Once this has been done, the cryptocurrency is created. This racing of solving a mathematical puzzle, and the first one to solve it getting the reward, is called the proof-of-work method of mining.

The more miners the more verifications of cryptocurrency transactions and therefore the more secure the network. To be a miner, it requires specialised computer hardware and software systems. As the popularity of cryptocurrencies increased, more miners join the network making more difficult for individuals to solve math problems. To overcome this, miners have developed a way to work together in pools which find solutions faster than individuals. Individual miners within these pools are awarded with cryptocurrencies proportionally to the work each of them provides in the mining pool.

4.4.2.2 *Trading/Exchanging*

This is the easiest way to obtain cryptocurrency through hundreds of exchanges where they allow trade with cryptocurrencies and fiat currencies. Cryptocurrencies can also be bought person to person, and these transactions are normally through the exchange platforms. Any purchase of cryptocurrencies should not be undertaken without properly understanding the market trends and all of the inherent risks.

Before selecting an exchange, a digital wallet must be chosen to store the cryptocurrencies purchased. Digital wallets can be in desktops, smartphones and other mobile devices, servers in the cloud, etc. These digital wallets will track the total value of the cryptocurrencies allowing to conduct, monitor and keep track of the history of various transactions. We might think of an e-wallet as a sort of email system where instead of sending and receiving messages, we can send and receive digital money.

The first cryptocurrency which started trading in 2009 was Bitcoin (BTC), but according to https://coinmarketcap.com/all/views/all/ in the end of April 2019, there were around 2140 cryptocurrencies often referred as Altcoins, the top 20 of which in terms of their market capitalisation can be found in the Appendix section.

4.4.3 *Popular Cryptocurrencies*

Although in terms of market capitalisation the list of top 20 cryptocurrencies changes continuously and dynamically, the top 5 most popular cryptocurrencies, used throughout the last recent years in most of the darknet markets and still remain in the above top 20 list, follow in the next paragraphs. Their charts with their market values and price variations throughout these recent years are illustrated in the Appendix section.

4.4.4 *Bitcoin (BTC)*

[1]Bitcoin was the world's first and largest cryptocurrency which operates using peer-to-peer open-source technology in which everyone can take part with no specific owner and central control. Its founder is known to the world by the pseudonym Satoshi Nakamoto, and many experts have called it 'digital gold'.

[1] https://en.bitcoin.it/wiki/Promotional_graphics

At the time of writing, Bitcoin's market capitalisation is roughly $ 92.5 billion. The first time Bitcoin grew increasingly prohibitive was in the 1-year period between December 1, 2016, and the end of November 2017, when the BTC/USD skyrocketed from $746 to over $11,000. Its total supply will be strictly limited to 21 million Bitcoins, more than 16 million of which have been mined by people who solve complex mathematical algorithms in order to verify transactions and release blocks of Bitcoins (Fortrade 2018).

Bitcoin makes use of the SHA-256 algorithm which is generally considered as one of the most complex algorithms, while at the same time allows a greater degree of parallel processing. Consequently, Bitcoin miners in recent years have utilised increasingly sophisticated methods for mining Bitcoins as efficiently as possible. Nowadays, the most dominant Bitcoin mining method consists of the use of application-specific integrated circuits (ASICs). These are hardware systems which, unlike the simple CPUs and GPUs which came before them, can be tailor-made for the sole purpose of mining Bitcoins. The practical consequence of this innovation has made Bitcoin mining increasingly out-of-reach for the everyday user (Fernando 2019).

Having the advantage of the first and longest presence with the largest number of users, Bitcoin's network effect is much higher than the other cryptocurrency networks making Bitcoin one of the most popular networks for the people to join. It's worth mentioning that Bitcoin is used in almost all darknet markets and vendor shops in the Dark Web.

4.4.5 *Ethereum (ETH)*

[2]Ethereum is the name of a blockchain company that has created the digital token Ether. Ethereum and Ether are used interchangeably referring to the cryptocurrency (Kharpal 2017). Ethereum, launched in June 2015 by its inventor Vitalik Buterin, is a relatively new, developing and promising cryptocurrency.

The annual rate of the currency's issues from mining is 18 million ETH. Many consider it as a replacement for Bitcoin. While the Bitcoin blockchain has tended to be used for consumer payment transactions, the Ethereum blockchain technology is intended to be used by the corporate

[2] https://worldvectorlogo.com/logo/ethereum

world. Its purpose is to provide businesses with the platform they need to build their services and products supporting smart contract applications that can automate complex physical and financial supply chain procedures and compliance processes involving multiple parties. The list of companies, including Microsoft, Intel, BP, JPMorgan and Cisco, is constantly increasing most of them becoming members of the Enterprise Ethereum Alliance (EEA), established in February 2017 seeking to use blockchain technology to run smart contracts at Fortune 500 companies (Ainger 2017).

While Bitcoin users tend to be politically and economically conscious as the counter to central banks and big governments, Ethereum users are less ideologically driven with its community focusing on the technology's future business and financial applications (Hay 2018).

As of December 20, 2017, when its value was skyrocketed within only a few months from $100 to $823, the market capitalisation of Ethereum was over $79 billion, making it the second largest digital currency in the world, after Bitcoin, with the expectations to continue its upward course.

4.4.6 *Litecoin (LTC)*

[3]Litecoin was released on GitHub on October 7, 2011 by Charlie Lee, a former engineer of Google. It was intended to be the silver to Bitcoin's gold (Bajpai 2019), according to its founder's vision to act as complementary rather than competitive to Bitcoin. It is considered as the younger brother of Bitcoin although the number of available Litecoins is 84 million comparatively to the availability of 21 million Bitcoins.

Since the start of 2017, Litecoin has risen 7291% against Bitcoin's 1731%. One of the reasons for this rise is that some investors are becoming more adventurous; they may also think the price of Bitcoin is overdone and are seeking other investment opportunities using alternative lower price and newcomer cryptocurrencies (Shen 2017).

Litecoin targets merchants who need a large volume of small transactions to be processed relatively quickly. The transaction processing speed of Litecoin network's long-term average transaction confirmation time is roughly 2.5 minutes making it faster and thus more attractive than Bitcoin which is around 10 minutes per transaction. This means that a seller needs

[3] https://worldvectorlogo.com/logo/litecoin

to wait for transaction confirmation in Bitcoin four times longer than in Litecoin. Another technical type of difference between Bitcoin and Litecoin is the different cryptographic algorithms they use. Bitcoin makes use of the long-standing SHA-256 algorithm, whereas Litecoin makes use of a comparatively new algorithm known as scrypt (Fernando 2019).

One of the main issues with Litecoin is that many exchanges trade it only for Bitcoins and not for USD or EUR in which case the only option is to buy Bitcoins and then exchange them to Litecoins. However, there are some exchanges that allow you to buy Litecoins directly (Beigel 2018).

4.4.7 Monero

[4]Monero (XMR) is an open-source, anonymous and decentralised relatively new and uprising cryptocurrency, created in April 2014, based on CryptoNight PoW hash algorithm coming from the CryptoNote protocol. Monero makes faster transactions using mining algorithm which does not favour application-specific integrated circuits (ASICs), but mining can be done using normal computers with any CPU or GPU, because it was designed to attract more 'little' nodes rather than rely on a few farms and mining pools. Unlike most cryptocurrencies, Monero has always-on privacy features applied to its transactions. Therefore, when someone sends Monero to somebody else, you can't tell who sent it to whom, and you can't backtrack their transactions since Monero's cryptography shields the sending and receiving wallet addresses keeping them unknown (Arntz 2017).

Unlike most of the digital coins, Monero is also untraceable using a special technology called 'ring signatures' which shuffles users' public keys in order to eliminate the possibility to identify a particular user. Due to its privacy, Monero is fungible. Units of Monero cannot be blacklisted by vendors or exchanges due to their association in previous transactions. Monero is also resistant to blockchain analysis due to its unlinkability feature, which was achieved by its protocol that generates multiple one-time public addresses that can only be gathered by the message receiver and hardly analysed by foreigners inside the block explorer (Monero 2019).

[4] https://www.getmonero.org/

4.4.8 *Dash/Darkcoin*

[5]Dash was created in January 2014 by Evan Duffield and it was originally called ‘Darkcoin’ which then in March 2015 was finally rebranded to ‘Dash’, which stands for Digital Cash (Bajpai 2019).

Dash has introduced a feature called InstantSend, which automatically freezes funds as soon as the money is sent, before the transaction has been confirmed. This feature prevents the double spending problem to occur like in most cryptocurrencies when a buyer conducts two transactions almost simultaneously; the funds used in the first purchase are still available for the second one, and thus it is possible that a buyer will spend more than he has available in his wallet. Since January 2017, Dash’s price is up 8000%, roughly six times the rise of Bitcoin, according to CoinMarketCap.com. Unlike Bitcoin and other cryptocurrencies, where miners do all of the work on the blockchain, Dash has a tier of superusers called ‘masternodes’ who perform key functions like deciding which projects get funded enabling private transactions on Dash. Due to this masternode structure, since power is concentrated in the hands of some users, some critics say it’s not truly a decentralised network (Kauflin 2017).

4.5 Active Markets

The Dark Web market sites are actually classified into two major groups: the marketplaces and the vendor shops which are the suppliers of the dark markets having their own sites and stores either separate or embedded within the marketplaces. Most of the marketplaces apart from providing trading services also act as forum spaces for their users’ community.

According to the three most popular websites dedicated to Dark Web information, the DeepDotWeb (2019b), the Dark Web News (2019) and the Darknet Markets (2019) where all dark markets are listed with their corresponding web links and users’ reviews, 15 marketplaces and 14 vendor shops were found active until early May 2019 and around 175 inactive as analysed in Fig. 4.1.

It’s also worth mentioning that many markets’ status, although active, can go down or up from time to time. Visiting the above darknet market

[5] https://www.dash.org/

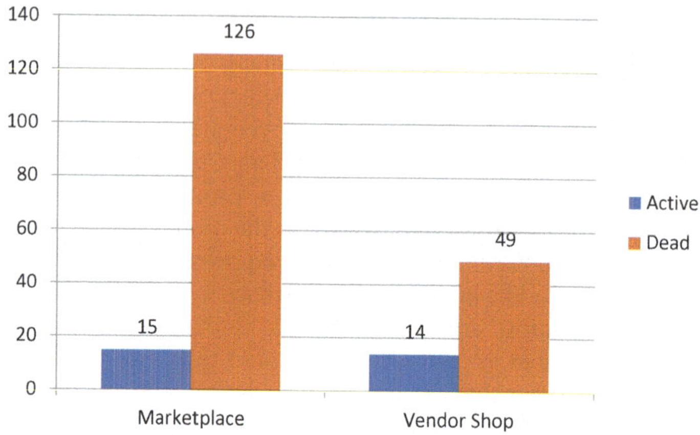

Fig. 4.1 Active and dead marketplaces and vendors shops

websites, you can find the uptime % status of the active markets, their current availability status with the sites which are down, marked in red. By the time someone reads these paragraphs, the figures of active and dead markets and vendors are very much likely to change. Reddit[6] is another good place on the web to start a research before using any hidden marketplace.

As it was observed throughout the 2-year research (2017–2019), the average number of active darknet markets and vendor shops at any period of time remained more or less constant to around 30 altogether in total, out of which though a small number of the same markets remained active in the list. Around 45% of the active markets are for specific language and countries, most of them for Russian, Italian and French markets and territories.

Taking into account that most of the marketplaces sell not just one but various products and services, the percentages of active marketplaces per type of goods and services, which are currently being sold in the active darknet markets, are illustrated in Fig. 4.2.

Thirteen out of the 15 active marketplaces make the 87% of those markets dealing with drugs; 5 out of the 15 active marketplaces make the 33% of those markets dealing with frauds and forgeries and also the same for guides and tutorials and so on.

[6] https://www.reddit.com/r/DarknetMarkets/

A short description follows for each of these most popular and longest-lasting active markets.

4.5.1 *Wall Street Market*[7]

An escrow and multisig trusted type of market, making use of Bitcoins, it is considered as one of the most good-looking, innovative and modern marketplaces for physical and digital goods with relatively few vendors most of its goods and service listings consisting of drugs, fraud and guides. Account registration process is very simple and fast, and although to become a seller or a vendor you have to apply for seller account, there is no fee mentioned in the site.

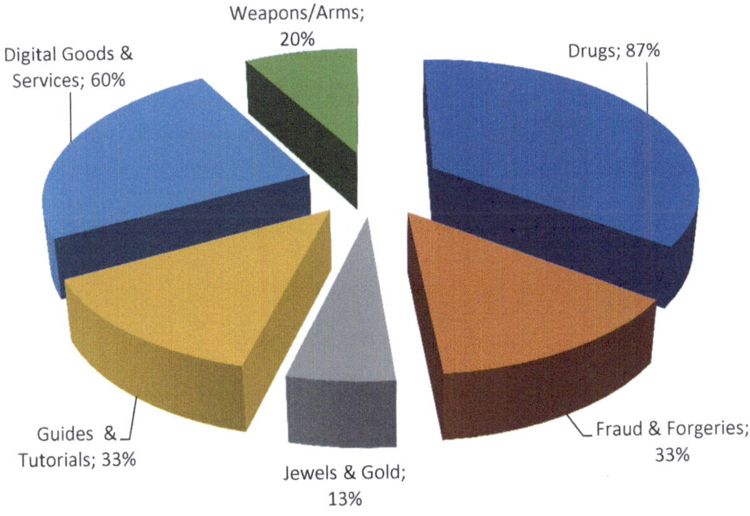

Fig. 4.2 Products/services mix in the 16 active darknet markets

---

[7] http://wallstyizjhkrvmj.onion/signup

4.5.2 *Dream Market*[8]

It's one of the world's largest and long-lasting darknet markets which has been around since November/December 2013 popular for its security and reliability and its simple user interface. The registration process is short allowing the user to get an account without revealing personal details. Its listings of goods and services include recreational drugs, drug paraphernalia, banned digital products, hacking and falsification services, counterfeits and weapons with their deliveries having some small delays. It is an escrow market using only Bitcoins which can be stored to the open-source Electrum wallet. Dream Market as a buyer-oriented market gives special focus on vendor evaluations by including not only its own market ratings but also ratings from other major marketplaces. There is also an official Dream Market forum.

4.5.3 *Nightmare Market*[9]

Nightmare Market Is a new uprising multi-sig darknet marketplace, offering new exiting features, very active support and enhanced buyer/seller security, with direct deposit supported. Accepted cryptocurrencies are Bitcoin, Bitcoin Cash, Litecoin, Monero, Zcash and Dash.

4.5.4 *Empire Market*[10]

Empire Market is a multisig market launched in early 2018 and is modelled after AlphaBay Market which was seized in mid-2017. The new market, as mentioned on the website, exists in memory of AlphaBay Market. The categories of the market include drugs, frauds, counterfeit items, digital products, guides and tutorials, software and malware, etc. The 'drug' category is the biggest that can be found on this marketplace; the market has also a forum

[8] http://4mtu5pl6yp3fmvny.onion/?ai=1675
[9] http://nightmareocykhgs.onion
[10] http://empiremktxgjovhm.onion

open to any member. Depending on the category one is involved in, there is an appropriate section to post. Accepted cryptocurrencies are Litecoin, Monero and Bitcoin.

4.5.5 Point/T•chka Free Market[11]

T•chka rebranded as Point Marketplace is a Russian marketplace dominated by drug listings which has been in operation since January 2015. It is not as big as other major marketplaces on the Dark Web, but it is well organised and good looking. Registration process is simple and quick with a choice to open a buyer or a vendor account. Accounts can be upgraded to premium with 0.2 BTC for more privileges. Its listings consist of drugs, prescriptions, digital goods and services and guides and tutorials. T•chka gets a 5% commission for sales through the site reduced to 2% for premium accounts. T•chka uses a traditional escrow system where the payment sent for the transaction will be held in escrow for up to 10 days and once the item is received the transaction will be finalised for the payment to be released to the vendor from escrow.

4.5.6 Silk Road 3.1[12]

Silk Road was originally founded in 2011 by **Ross William Ulbricht**, arriving as one of the first and most popular existing darknet markets at the time. Since then and after its founder was arrested back in 2013, the market was closed. The second version of the Silk Road was Silk Road 2.0 which eventually failed after around a year or so. The Silk Road 2.0 creator was also busted. This was a major disappointment to users whose lost money from the seizure. As a continuation of the second, the third version (Silk Road 3.0) was viewed as a scam. Then Silk Road 3.0 emerged as a copy of the original Silk Road but with much better security and a new team, which was behind Crypto Market. Silk Road 3.0 was taken down in 2017 by its administrators, and a new

[11] http://tochka3evlj3sxdv.onion
[12] http://silkroad7rn2puhj.onion

darknet market emerged officially launched as Silk Road 3.1. The product listings are mainly drug-related items.

4.5.7 *Cannabis Growers and Merchants Cooperative (CGMC)*[13]

It's a private invite-only cannabis market operating since June 2016. It has one of the most complicated displays full of logos and images. CGMC has neither many vendors nor many listings, most of them being concentrates and flowers. Vendors are carefully screened and selected. Bitcoin is the currency used in this market, and in order to pay for transactions, funds should be in your Bitcoin account. Payments on CGMC can be held in escrow through multisig payments and also by direct-type method.

4.5.8 *Berlusconi Market*[14]

It has a familiar look with Hansa market with intuitive and simple interface. The registration process is quick and easy. Payments are done with Bitcoins and Litecoins not through the account's wallet but directly from the user's own wallet to the seller's wallet. There is a large variety of categories and large number of listings in each category including drugs and chemicals, weapons, frauds and counterfeits, guides and tutorials, jewels and gold and digital goods and services. The vendors' pages provide a lot of useful information for the convenience of the buyers to make their selections. Feedback ratings, number of orders processed, date of first sign-up and last active status are some of these information on their profile pages. Purchasing is done by placing an order.

[13] http://cgmcoopwhempo6a5.onion/

[14] http://berluscqui3nj4qz.onion

4.5.9 *The Majestic Garden*[15]

It is a forum presenting and exchanging psychedelics-related information rather than a native marketplace with no categories and listing pages available. Nevertheless, there is a set of threads labelled vendor information and reviews which facilitate trading of psychedelics through P2P transactions or direct buyers to the above labelled vendors in other darknet markets. It has no wallets, no need for deposits and no fees for vendor accounts. It requires a simple account registration which is a forum and not a buyer/seller account with only three fields to fill in: username, password and an email which is recommended to be a separate one from those of regular everyday use.

4.5.10 *Hydra*[16]

It is a Russian darknet market which facilitates but isn't directly involved in the trading transactions between buyers and sellers; it supports though dispute resolution. It provides a quick registration process with a short form to fill in. Vendors in order to sell their goods have to create their store in the marketplace. Hydra specialises in recreational drugs and narcotic substances. The purchase process is based on order placement and not on buy now, which makes the buyer wait for the vendor to contact and confirm the order before it is initiated. Bitcoin and QIWI are the only options for paying transactions mentioned on the informational pages of Hydra.

4.5.11 *WayAway*[17]

WayAway is the oldest darknet market running since 2009, located in the past in the traditional web. It's a Russian forum-based marketplace, similar to RuTor market, and popular for specialising in recreational drugs and research chemicals. It focuses on Russian

[15] http://talismanrestz7mr.onion/
[16] http://hydraruzxpnew4af.onion/
[17] http://wayawaytcl3k66fl.onion/

market with no international section. There is no clear categorisation of products; emphasis is given to vendors or sellers instead. The account set-up process is easy and quick. All sale terms are defined by the vendors, and they are agreed directly with the buyers without the market's involvement. Payments in WayAway are made in Bitcoins.

4.5.12 RuTor[18]

It is a Russian-only darknet forum, but it also serves as a darknet marketplace for various types of products and services. There is a fee for sellers' profiles. The major categories for products on RuTor are drugs, weapons and various services with limited number of posts and vendors. It accepts deposits in Bitcoins for purchases using not an escrow system but a Bitcoin wallet address provided not by an automated system but by the administrator after buyer's request.

4.5.13 Cannazon[19]

It's a multisig marketplace dedicated to *Cannabis*. Accepted cryptocurrencies are Bitcoin and Monero.

4.6 Active Vendor Shops

These sites are not markets; they are individual vendors who operate their own sites, and they supply their goods to several darknet markets. According to the information collected till early May 2019, the 14 active vendor shops in the Dark Web are the following:

4.6.1 CharlieUK[20]

CharlieUK is a cocaine vendor from UK operating since 2013 on Abraxas, AlphaBay, Crypto and Dream Markets.

[18] http://rutorzzmfflzllk5.onion/
[19] http://cannazonceujdye3.onion
[20] http://eeyovrly7charuku.onion/

4.6.2 *DutchDrugz*[21]

DutchDrugz is an old-time vendor selling on several markets psychedelics and others.

4.6.3 *RechardSport*[22]

RechardSport is a professional wholesaler in China, specialised in sportswear.

4.6.4 *ElHerbolario*[23]

ElHerbolario is a vendor from Middle Earth – selling cannabis products. Users can discuss anything about darknet marketplaces and drug-related topics.

4.6.5 *Gammagoblin-Pushing Taboo*[24]

This is a personal shop opened by one of the Dark Web's most famous vendors, Gammagoblin, a vendor of psychedelic substances since SR1. It is star rated by LSD Avengers community.

4.6.6 *The Church-Jesus of Rave(JOR)*[25]

This is a private vendor shop of Jesus of Rave – old-time vendor since 2013 selling LSD and MDMA.

[21] http://dutchdr5gsol4dde.onion
[22] http://rechardsp4x6tdrh.onion/
[23] http://elherbotsiddarol.onion/
[24] http://pushingtabu7itqj.onion
[25] http://artsmankindxgcv5.onion/

4.6.7 *ToYouTeam*[26]

ToYouTeam is a vendor since 2012 selling multiple products.

4.6.8 *The French Connection*[27]

The French Connection is a vendor shop selling H, XTC, meth, etc. to Mr. Nice Guy or Nucleus.

4.6.9 *GlassWerkz*[28]

GlassWerkz is a vendor shop created by an Agora vendor who opened his own vendor shop. The shop was down because of an unknown reason.

4.6.10 *EU Cocaine*[29]

As its name states, it sells only cocaine. They ship worldwide and have personal technology for delivery, which depends on the size of the order accepting only Bitcoins.

4.6.11 *Cocaine Market*[30]

This is a vendor shop where only cocaine can be bought. The product can be bought only by Bitcoins.

---

[26] http://tytbeta57rw2onit.onion/
[27] http://abyssopyps3z4xof.onion/
[28] http://glasvyhbf444airs.onion
[29] http://cocahze7fqy4qwwx.onion
[30] http://cocain2xkqiesuqd.onion

4.6.12 *ChemSpain*[31]

They provide domestic and international shipping from Spain, accepting only Bitcoin payments. From this vendor shop GBL, RC, Benzos, GHB and Bulk are the products mainly sold.

4.6.13 *MaghrebHashish*[32]

MaghrebHashish Store

MaghrebHashish is opened by a hashish and weed seller from Crypto Market and AlphaBay named Maghreb.

4.6.14 *Mushbud*[33]

Mushbud sells weed and psychedelics to markets since the original Silk Road as well as SR2, Sheep, BMR, Agora and briefly at Abraxas.

4.7 Dead and Scam Markets and Vendors: The Ten Most Popular

One hundred and twenty six marketplaces and 49 vendor shops were found dead/scam until mid-2019. Some of these markets are going up and down from time to time, and they are considered temporarily as scam/dead although they are still in operation. A small description of the 12 most popular ones are listed below.

4.7.1 *AlphaBay*[34]

a AlphaBay Market

AlphaBay is an English commercial darknet market one of the largest darknet market in the world launched in December 2014 with more than 200,000 users and 40,000 vendors prior to its takedown. There were over 250,000 listings with a wide range of categories including fraud, recreational drugs, chemicals, counterfeit products, digital goods, weapons, carded electronics and

[31] http://chemspain7iw2zby.onion
[32] http://mghreb4l5hdhiytu.onion
[33] http://kpj3orlxmfs6yqah.onion
[34] http://pwoah7foa6au2pul.onion

appliances, carded clothing, jewellery, gold, software, security and hosting solutions, various services as well as guides and tutorials. AlphaBay supported two cryptocurrencies: Bitcoin (BTC) and Monero (XMR).

4.7.2 *Agora*[35]

After Evolution closed in an exit scam in March 2015, Agora replaced it as the largest drug market on the darknet and has proven itself to be ultra-reliable. It launched in 2013 and shut down in August of 2015.

4.7.3 *Evolution*[36]

evolution

Launched in January 2014, Evolution was the second or third largest market by trade volume offering multisig and normal escrow services. It earned a reputation for its security and its reliability, with a high uptime rate. The shutdown in March 2015 was found to be an exit scam, with the site's operators stealing approximately $12 million in *Bitcoins* it was holding as escrow.

4.7.4 *Hansa Market*[37]

HANSA

It was not as big as the other leading marketplaces on the Dark Web. Nevertheless, it featured thousands of listings in a number of categories including drugs, jewellery, frauds and counterfeits, guides and tutorials and digital goods and services.

[35] http://agorahooawayyfoe.onion
[36] http://k5zq47j6wd3wdvjq.onion
[37] http://hansamkt2rr6nfg3.onion/dashboard/settings/

4.7.5 *Outlaw Market*[38]

It was one of the oldest darknet markets having been founded back in 2013 when Silk Road 1 was still up. It was developed with emphasis on security and on keeping the project away from open-source platforms for less vulnerability to exploits. Outlaw Market dealt in a variety of illegal products including weapons, drugs and data dumps. It went offline on May 16, 2017. According to a message posted on the homepage, the site was shut down due to a hack. It also stated that the platform's wallet had been stolen. The fact though that Outlaw Market was developed with an emphasis on superior security features gives reduced chances for hacking making many users to believe that it was an exit scam orchestrated by the site's administrators.

4.7.6 *Aero Market*[39]

Aero Marketplace and Forums was a new generation of darknet black market, comprising of the latest in server security technology and platform development from a highly skilled team. It provides a user-friendly and very clean looking interface for both buyers and vendors that makes operations much easier. Registration process was very simple. Registered users could be both buyers and vendors with one account. Vendors, in order to open their store and sell, had to pay a one-time fee which varies according to the plan they would select. More than 60% of the items listed in this market were drugs. Digital goods and services, frauds and counterfeits, guides and tutorials, jewels and gold constituted the rest of the item listings. The transaction types were either escrow or 'finalise early' and also multisig supported with the facility for use of both Bitcoin and Monero. Late November 2017 the market went down with warnings for suspected scam.

[38] http://outfor6jwcztwbpd.onion/

[39] http://aeroguckedxcpwoa.onion/

4.7.7 *Libertas Market*[40]

It is a Monero-only darknet market with simple design and interface having the green as the distinctive colour of its website theme. Registration process takes sometime due to the long registration form which has to be filled. A variety of few listings are grouped in the following main categories: drugs, forgeries, jewels and lab supplies.

4.7.8 *TheRealDeal Market*[41]

TheRealDeal was shut down in November 2016. This darknet market was known to be part of the cyber arms industry selling code and zero-day software exploit. As the creators claimed in an interview given to DeepDotWeb, the marketplace was created to give a direct response to all those Dark Web sites which do not actually have anything of value to sell and are just scams. More than 40% of the listings on the site were exploit and fraud related. The drug-related listings, however, had a share of more than 50%.

4.7.9 *Mercado Negro*[42]

Black Market as its name means in Portuguese is an initiative that unites Brazilians and the Netherlands aiming at the freedom of consumer goods. It is developed in one of the most secure platforms in the world, aiming at security, facilitation and usability. It uses Monero as a form of payment.

[40] http://libbyxh6som2twgp.onion
[41] http://trdealmgn4uvm42g.onion
[42] http://mercadouipzapomr.onion/

4.7.10 *Italian Darknet Community*[43]

IDC, similar to the Majestic Garden, is a darknet forum mainly created for Italian users which apart from providing a secure platform for discussing various topics it also allows members to post product and service offers. IDC has a membership system with different types of accounts such as base, seller and elite accounts with different fees applied after the quick registration process. The market section consists of drugs and forerunners, documents and fake money, carding, hacking and coding, electronics and other stuff. There is also an international section for French and English with different items mainly cloned and fake cards, accounts and counterfeit items. IDC has an escrow system and uses Bitcoin.

4.7.11 *Valhalla (Silkkitie)*[44]

Valhalla (Silkkitie)

It was introduced in October 2013 originally called Silkkitie which used to focus on Finish market. It was rebranded after 2 years of operations and served the global Dark Web. Valhalla had an invite method of registration in order to control the growth of the market and to encourage members to invite other users. Most of its listings were drugs and narcotics-related products providing three payment methods: escrow/multisig/FE, escrow/multisig and multisig only.

4.7.12 *Rsclub Market*[45]

It was a relatively new and small escrow marketplace with limited vendors and listings which mainly consist of frauds and counterfeits, guides and tutorials and digital goods and services. Its features compete with those of other leading dark markets such as autoshop with good presentation of listings and a good account interface, seller and trust levels, a well-organised feedback system for buyers to evaluate sellers and a nice internal messaging platform. Bitcoin was

[43] http://2qrdpvonwwqnic7j.onion/

[44] http://valhallaxmn3fydu.onion/

[45] http://rsclubvvwcoovivi.onion/

the currency used in the market. Long session timeout, occasional 'unable to connect' errors and long downtimes were some of the cons observed in the market's reviews.

4.8 Policing the Dark Markets

The three investigation phases in the dark market policing include the strategic investigation, the identification and the prosecution of suspects. How successful the first strategic phase of Dark Web policing will be depends either on intrinsic characteristics of the technology and environment of the dark markets or on organisational and governance aspects of the broader ecosystem where law enforcement needs to operate. In the second phase of the identification of suspects, the goal can be that of surveillance of suspects to research their identity and responsibility in the crime. If a criminal investigation lead to a geographical area, for example, when a server hosting dark market services is located, a local operation is typically set up. Finally, in the third phase of the suspect's prosecution once an investigation successfully lead to an arrest, sufficiently solid proofs and evidences need to be in place to bring the case to court (MEDIA4SEC-TNO(Serena Oggero) 2017).

The coordinated and efficient actions of European and American law enforcement agencies against the illegal trading in the Dark Web led to the takedown of several online markets and their operators behind them. After a numerous investigation and coordinated month efforts between FBI, US Drug Enforcement Agency (DEA) and the Dutch National Police with the support of Europol, two of the most popular marketplaces, AlphaBay and Hansa, were shut down on July 20, 2017. Europol with the help of an Internet security company, the Bitdefender, and the support of partner agencies in the Netherlands, Germany and Lithuania provided Dutch police the means to lead the investigations, locate the infrastructure of Hansa Market, arrest its two administrators, seize its servers and take over the Hansa marketplace on June 20, 2017 till it was eventually shut down on July 20, 2017. This taking over control operation of Hansa by the National High-Tech Crime Unit of the Netherlands allowed them to collect significant information for sellers and buyers, monitor the users' activities without their knowledge and during the same period shut down also the AlphaBay Market under a coordinated operation, called Bayonet. In the mean time between the taking down of AlphaBay and of Hansa Market, a large number of new members displaced from AlphaBay to

Hansa enabling the Dutch High-Tech Crime Unit to record and identify them disrupting their criminal activities on Hansa (Europol 2017a, b).

One of the most successful operations of law enforcement agencies was the arrest of Ross Ulbricht, the founder of Silk Road, the first and the most popular market in the recent history of Dark Web. After a series of more than 1-year investigations, authorities managed to arrest him in 2013. He is now serving two lifetime sentences in jail.

In November 2014 more than 20 darknet markets and more than 400 of scam websites selling illegal items including drugs and weapons were shut down by the authorities with an operation known as Onymous Operation. The operation involved the police forces of 16 European countries and the USA leading to 17 arrests, including Blake Benthall who was said to be behind Silk Road 2.0 one of the most notorious markets in the buying and selling of illegal drugs (Wakefield 2014).

In December 2016, two arrests were made by the Slovenian law enforcement authorities, with the support of Europol, of two suspects accused of selling various firearms and weapons including automatic rifles, hand and smoke grenades as well as ammunition through a prominent darknet market using Bitcoin (Aliens 2017).

According to the Europol's press release on September 9, 2016, a small number of cases have shown LEAs that money laundering and terrorism financing can easily take place with the use of digital currencies inside virtual environments, offering high-level anonymity and low-level risks comparatively with those associated with real-world money laundering and terrorism financing activities. For this reason Europol, Interpol and the Basel Institute on Governance joined their forces on the above date and formed a partnership for a working group on money laundering with digital currencies. Their aim is to gather, analyse and exchange non-operational information from the digital currency use, to organise workshops and meetings for LEAs and institutions to increase the investigation capacities in crimes in which cryptocurrencies get involved and last but not least to create a network of practitioners and experts in this field, who can collectively establish best practices and provide assistance and recommendations inside and outside the working group (Europol 2016).

4.9 Future Trends, Challenges and Opportunities

4.9.1 *Multifunctional Decentralised Markets*

In the forthcoming years, the trend for the Dark Web markets is to move to all-in-one privacy-oriented social platforms. UMBRA is one of these privacy platforms which is currently being built by the developers of the ShadowCash cryptocurrency. The platform is a decentralised anonymous commerce, currency and communication space with multiple functions such as a P2P marketplace (ShadowMarket), secure payments and transfers (ShadowCash and ShadowSend), cryptocurrency wallet management and encrypted chat (ShadowChat). Another under development platform is Komodo (Brown 2016).

Platforms like these aim to attract the suppliers and customers of tomorrow's darknet markets to develop communication and transactions providing them an all-in-one place of multiple functions with strong privacy, security and ease-of-use features.

On the other hand, new decentralised marketplaces are likely to overcome the weakness and vulnerability of being hosted in a specific location. These localised markets bring the sellers and buyers closer to interact more directly in their own language cutting the intermediaries and arranging goods deliveries at local level avoiding international mail postings (SOCTA-Europol 2017).

4.9.2 *Less Vulnerable Systems*

Dark markets are becoming more and more technically complex for security enhancement purposes creating a trade-off though between the complexity for security vs the user accessibility. Although the technology enables the development of more secure and less vulnerable systems, the human mistakes still remain as the main weaknesses for the criminals but at the same time as the best opportunities for the law enforcement agencies to trace and identify them.

4.9.3 *Transformation of Cryptocurrency Networks to Business Service Platforms and Alliances*

The original Bitcoin design and the rest of the cryptocurrencies which were developed based on the blockchain technology have created the inspiration to use this technology not only for digital money but for other applications related to production, commercial and financial chain of transactions. These new applications based on blockchain technology gave the chance to some cryptocurrency networks like Ethereum to go one step further and create new growing corporate platforms and alliances such as the Enterprise Ethereum Alliance (EEA) already mentioned above in this chapter.

4.10 Concluding Remarks

The increasing and continuously evolving complex Internet and mobile technology in the last decade facilitated the operation of Dark Web services and darknet markets for trading of illegal products and provision of illicit services enabling in turn the development of underground economy through transactions carried out between buyers and sellers based on new types of digital currencies, called cryptocurrencies.

The total number of most popular dark markets that appeared in this decade was around 130, but only 12% of this figure, e.g. around 15 markets, remain active nowadays with a trend to further reduction. The same applies more or less to the vendor shops. This low figure of active marketspaces is explained due to three main occurring facts:

1. New market developers are not very well aware of the rapidly evolving new technology with the new security- and vulnerability-associated issues to apply, and thus they fail fast.
2. Darknet market owners and administrators as soon as their markets grow get the money gathered in the market, and they disappear proceeding with sudden exit scams.
3. LEAs' effective coordination and interventions cause more and more shutdowns of darknet markets.

Regarding the cryptocurrencies issued so far, although more and more new currencies are appearing everyday with some of them having tremendous and rapid values and price increases in a very short fraction of time, only a few of them are still used in the darknet markets as more trusted and longest in use such as the Bitcoin, Monero, etc. Due to the fact that prices of cryptocurrencies rise fast, the cryptocurrency market started becoming a stock exchange type of gambling place where people started investing in new currencies expecting high gains from buying and selling their currencies without actual trading in goods.

Multifunctional, decentralised and less vulnerable systems are the new trends of darknet markets, whereas the trends of the blockchain technologies on which cryptocurrencies are based on tend to be expanded in other applications and platforms in the business world bringing corporations together in production and commercial alliances.

One of the main conclusions from the Dark Web stakeholders' analysis that came out from the workshop in Dark Web, held as part of the European research project MEDI@4SEC, on September 26, 2017 in The Hague, Netherlands, is that the whole dark market ecosystem revolving around the business value chain acquires a potential responsibility in reducing crime. This includes, apart from law enforcement agencies, private sector actors such as fintech and financial services, logistics providers, Internet service providers and citizens at large. Proactive collaboration between the private and the public sector, coordinating operations and stimulating a collaborative sharing culture between law enforcement agencies at local, regional, national, European and international level, reducing double efforts by sharing common repository of data, information knowledge, tools, practices and methods, and a common legislation framework in place, were some of the main recommended actions (MEDIA4SEC-TNO, Serena Oggero 2017).

References

N. Ainger, *Bigger than Bitcoin? Enterprise Ethereum Alliance Grows in Size.* [Online] (2017), Available at: https://www.cnbc.com/2017/05/23/bigger-than-bitcoin-enterprise-ethereum-alliance-grows-in-size.html. Accessed 4 May 2019

C. Aliens, *Europol Links Darknet Markets and Terrorism* (DEEP.DOT.WEB, s.l., 2017)

P. Arntz, *How Cryptocurrency Mining Works: Bitcoin vs. Monero.* [Online] (2017), Available at: https://blog.malwarebytes.com/security-world/2017/12/how-cryptocurrency-mining-works-bitcoin-vs-monero/. Accessed 4 May 2019

P. Bajpai, *The 6 Most Important Cryptocurrencies Other Than Bitcoin.* [Online] (2019), Available at: https://www.investopedia.com/tech/most-important-cryptocurrencies-other-than-bitcoin/. Accessed 4 May 2019

O. Beigel, *How to Buy Litecoin with Paypal, a Credit Card or a Bank Transfer.* [Online] (2018), Available at: https://99bitcoins.com/buy-litecoin-with-paypal-credit-card-instantly/. Accessed 4 May 2019

B. Brown, *State of the Dark Web* (AKAMAI, s.l., 2016)

Dark Web News, *Dark Web News-Market List*. [Online] (2019), Available at: https://darkwebnews.com/dark-web-market-list/. Accessed 4 May 2019

Darknet Markets, *Best Deep Web Market Links, List and Reviews*. [Online] (2019), Available at: https://darknetmarkets.co/. Accessed 4 May 2019

DeepDotWeb, *Escrow Markets*. [Online] (2019a), Available at: https://www.deepdotweb.com/marketplace-directory/categories/marketplaces/. Accessed 4 May 2019

DeepDotWeb, *Home Page*. [Online] (2019b), Available at: https://www.deepdotweb.com/. Accessed 4 May 2019

DeepDotWeb, *Invite Markets*. [Online] (2019c), Available at: https://www.deepdotweb.com/marketplace-directory/categories/invite-markets/. Accessed 4 May 2019

DeepDotWeb, *Multisyg or Trusted Markets*. [Online] (2019d), Available at: https://www.deepdotweb.com/marketplace-directory/categories/multisig-and-trusted/. Accessed 4 May 2019

Europol, *Money Laundering with Digital Currencies: Working Group Established*. [Online] (2016), Available at: https://www.europol.europa.eu/newsroom/news/money-laundering-digital-currencies-working-group-established. Accessed 4 May 2019

Europol, *Massive Blow to Criminal Dark Web Activities After Globally Coordinated Operation*. [Online] (2017a), Available at: https://www.europol.europa.eu/newsroom/news/massive-blow-to-criminal-dark-web-activities-after-globally-coordinated-operation. Accessed 4 May 2019

Europol, *Serious and organised crime threat assessment (SOCTA)-Crime in the age of technology* (Europol, The Hauge, 2017b)

J. Fernando, *Bitcoin Vs. Litecoin: What's the Difference?* [Online] (2019), Available at: https://www.investopedia.com/articles/investing/042015/bitcoin-vs-litecoin-whats-difference.asp. Accessed 4 May 2019

Fortrade, *Bitcoin*. [Online] (2018), Available at: https://www.fortrade.com/products/btcusd/. Accessed 18 Mar 2018

S. Hay, *Bitcoin VS Ethereum: Cryptocurrency Comparison*. [Online] (2018), Available at: https://99bitcoins.com/bitcoin-vs-ethereum-cryptocurrency-comparison/?gclid=CjwKCAiAj53SBRBcEiwAT-3A2HDrfsPGM-HuH9pw33iRhogJxXBq6qogBpPAHLAqYnEyBGK26CjxQBoC7K4QAvD_BwE. Accessed 4 May 2019

J. Kauflin, *Dash Is Up 8,000% In 2017. Is this 'Darkcoin' a Better Version of Bitcoin?* [Online] (2017), Available at: https://www.forbes.com/sites/jeffkauflin/2017/12/22/dash-is-up-7600-in-2017-is-this-darkcoin-a-better-version-of-bitcoin/#784d1d9e1a02. Accessed 4 May 2019

A. Kharpal, *All You Need to Know About the Top 5 Cryptocurrencies*. [Online] (2017), Available at: https://www.cnbc.com/2017/12/14/bitcoin-ether-litecoin-ripple-differences-between-cryptocurrencies.html. Accessed 4 May 2019

MEDIA4SEC-TNO(Serena Oggero), *Workshop 3 Report: Policing the Dark Web* (MEDI@4SEC-EU Project, s.l., 2017)

Monero, *Why Monero.* [Online], (2019). Available at: https://monero.org/. Accessed 4 May 2019

L. Shen, *What Is Litecoin, and Why Is It Beating Bitcoin This Year?* [Online], (2017). Available at: http://fortune.com/2017/12/12/litecoin-bitcoin-price-2018/. Accessed 4 May 2019

SOCTA-Europol, *Europol.* [Online] (2017), Available at: https://www.europol.europa.eu/socta/2017/. Accessed 4 May 2019

M. Staalduinen, *Dark Drugs Markets After Operation Bayonet: Finito Basta?* [Online] (2017), Available at: https://dws.pm/2017/07/21/dark-drugs-markets-finito-basta. Accessed 4 May 2019

J. Wakefield, *Huge Raid to Shut Down 400-Plus Dark Net Sites.* [Online] (2014), Available at: http://www.bbc.com/news/technology-29950946. Accessed 4 May 2019

CHAPTER 5

TENSOR: A Solution to Dark Web Investigations

Tony Day, Kieran Dennis, and Helen Gibson

5.1 Introduction

A wide range of criminal activities, particularly those related to hate crime, radicalisation and terrorism, have an ever-increasing online component that needs to be considered, captured and investigated. Some of this crime is occurring in dark, hidden corners of the Internet or is conducted in encrypted private channels, where there is little that can be done to access it in real time. However, as with many types of crime, there is often a visible surface layer available to investigators that can provide vital intelligence. For example, a private group containing extremists can remain closed sharing their hate speech and radical ideals amongst themselves; but, many radical individuals want to spread their ideals through propaganda or even discuss or inspire others to carry out attacks and illegal activities. This is where such activity may manifest itself in the open, whether on message boards, forums, websites or social media platforms.

T. Day • K. Dennis • H. Gibson (✉)
CENTRIC, Sheffield Hallam University, Sheffield, UK
e-mail: t.day@shu.ac.uk; k.dennis@shu.ac.uk; h.gibson@shu.ac.uk

B. Akhgar et al. (eds.), *Dark Web Investigation*,
Security Informatics and Law Enforcement,
https://doi.org/10.1007/978-3-030-55343-2_5

Any and all of these methods of communicating and sharing information can be found across the surface, Deep and Dark Web, often available to all.

The effect of this sea change in criminal behaviour means law enforcement is having to go through radical changes in their investigatory processes to respond to this challenge and will continue to do so for the foreseeable future. Digital systems to support criminal investigations are becoming commonplace; however, aspects of the Dark Web still remain 'off limits' to investigators as they battle technical, regulatory and organisational challenges.

TENSOR is conceived as a system (see Akhgar et al. (2017) for an initial introduction to the vision for TENSOR) that is able to address many of these challenges by providing a mechanism from which investigators can capture intelligence from online sources including the Dark Web and convert it into structured data with a coherent data model (Sect. 5.3). The benefits of this structure are the ease with which the information within TENSOR becomes searchable, analysable (both through automated and investigator-led means) and relatable (links between multiple pieces of information are immediately evident). The purpose of this chapter is to walk through the rationale, key functionalities and potential use cases of TENSOR to distinguish it from existing OSINT or Dark Web investigation platforms, whilst demonstrating the added benefit TENSOR provides to the investigator.

The chapter proceeds as follows: firstly we illustrate the underpinning data model for TENSOR and combine this with a discussion of potential data sources, acquisition and extraction processes; secondly we review the core functionality of the underlying system for analysis; thirdly we cover the user interaction with TENSOR and how analysis is manifested in the intelligence dashboard interface; fourthly we review initial user feedback to motivate future work through a set of lessons learned.

5.2 The Role of TENSOR

TENSOR (retrieval and analysis of heterogeneous online content for terrorist activity recognition) focuses on enabling investigators to research and analyse content on publicly accessible virtual spaces where modern technologies are being exploited for nefarious and illicit causes. These virtual spaces occur on social media platforms, the surface web, or on hidden marketplaces or forums on the Dark Web. They represent anywhere that an individual can access relatively easily with simple technologies such as a web browser. Currently, many law enforcement agencies (LEAs) are bound to archaic approaches of searching, extracting and analysing this unstructured online content (Deloitte 2015), such as through screen-shots, which are only easily interpretable by humans and have severe limitations (e.g. see Feldman (2015)).

Recently, LEAs have had access to technologies that better support their capture of data from the web although such methods are still fraught with difficulties. First and foremost is the fact that no matter how advanced the technologies LEAs have access to the chances are criminals are already two steps ahead making use of new apps, communication channels and the ease with which they can switch aliases. Thus, law enforcement is always playing catch-up. Furthermore, the very tools that are also supposed to help them capture data also drown them in data with extensive configuration required to limit over-collection and a lack of analysis capability. This is often compounded by a misunderstanding of the role of analysts and a lack of training (Belur and Johnson 2018). Access to social media services for LEAs can also be switched off at a moment's notice which does not help when, after a crime, there are questions from the public and media if a perpetrator's intentions were there for all to view. Finally, most tools are not silver bullets, and with limited budgets, LEAs can only invest in licencing and training for the most broadly used.

TENSOR is a collaborative project between European partners funded by the European Commission's Horizon 2020 research and innovation programme. It aims to extend investigators' capabilities by replacing much of the repetitive and manual work required to understand, acquire, extract and analyse data from the ever-increasing range of open spaces on the web. TENSOR is particularly focused on addressing the proliferation of terrorist content posted online; however, many of the technologies developed within the project are applicable across a wide range of domains. TENSOR brings together a wide range of technical capabilities covering content acquisition and extraction, analysis and visualisation and intelligence management into a single space known as the TENSOR intelligence dashboard.

Using TENSOR, open-source law enforcement investigations can be both speeded and scaled up. Scaling up allows investigators to focus less on structuring and organising content and much more on identifying important patterns and connections between individuals, groups and the content they produce, share or interact with. Using a customised data model focused on connectivity, the investigator is able to work with heterogeneous data from the surface web, Dark Web and social media services, whilst only dealing with familiar concepts such as posts, profiles, pages, messages and hashtags. Taking advantage of a powerful content extraction capability, this content is indexed and categorised efficiently to allow searchability and analysis that simply cannot happen with the status quo of screenshots in Word documents. Using cross-platform fusion, TENSOR simplifies the identification of actors communicating over various systems and platforms with various aliases alongside the content they produce.

Crucially, TENSOR has been informed by law enforcement end users at every stage, taking advantage of the highly collaborative structure of a project between pan-European partners. Not only has this given TENSOR a clear strategic direction, it has also meant that ongoing feedback from law enforcement end users through three pilot and evaluation phases over the 3 years of the project has been directly incorporated during development. With such close participation between legal experts, law enforcement, academia and industry, TENSOR has utilised privacy by design (Langheinrich 2001) at a time when there has been a major shift in the way data protection is perceived and managed due to the General Data Protection Regulation (GDPR). TENSOR has had to be particularly mindful of these issues given its capacity to increase the scale and velocity of data collection and analysis of potentially sensitive information. Efforts have been made throughout the development to assure that any potential impacts on personal privacy have been outlined and mitigated, from the design, research and implementation perspectives using both law enforcement-focused legal and ethical controls but also through good security practices within development.

Considering how TENSOR aims to expedite typically menial content acquisition tasks so that highly skilled investigators can focus on their investigations, special emphasis has also been placed on protecting the chain of evidence. As a complex part of policing, it is difficult to replicate procedures for storing of physical evidence with digital evidence, especially that which has been obtained online (given that some standardisation already exists for typical digital forensics tasks relating to content acquisition from hardware). A common solution is digital hashing, employed regularly by law enforcement (Giova 2011), which helps to maintain the line of authenticity from evidence collection onwards. TENSOR introduces a further security measure of digital signatures integrated into comprehensive auditing system that tracks the introduction and subsequent modifications to each individual piece of content or entity. Protecting the integrity of the content with digital signatures can precisely record when and how the content was acquired and, crucially, cannot be modified without access to the secret key that created it.

5.3 Modelling Highly Connected Content

TENSOR has been designed around an abstract data model that emphasises the connectivity between content and the 'things' within it that replicates and amplifies the natural connections made by human investigators. This model is reflected in TENSOR's underlying secure storage repository – the hub of the system. At its core this data model is effectively a triplestore containing relationships structured in subject-predicate-object format. For example, consider the following triple: 'Actor A liked Post B'. In this triple, Actor A is the subject, 'liked' is the predicate, and 'Post B' is an object. The subject and the object both form entities within the system, whilst the predicate represents links. Triples are always directed, and the link can take many forms including 'follows', 'author of', 'shared' and many more.

Triplestores are often represented as a graph (or network) where objects are vertices (nodes) and predicates are edges (links). As predicates denote one-way relationships, a triplestore can be represented as a directed graph. This allows both for a representation of these links in a graph or network visualisation and for processing and exploration of the data, by 'walking' or 'traversing' the graph alongside the application of methods from both graph theory and social network analysis (SNA). For example, to find all friends of friends for a single actor (two steps away from our initial actor) would involve starting at the object representation of that actor, a node in the graph, and then for each of the nodes linked via a predicate of 'friend of', find all the nodes linked from them via the same predicate. These types of queries can be far more complex and as a result far more powerful, allowing for high-level analysis of the structure of the network.

5.3.1 *Artefacts, Entities and Links: The TENSOR Data Model*

Building on the idea of triplestores and entity-relationship models, the TENSOR data model is divided into three parts: artefacts, entities and links (Fig. 5.1).

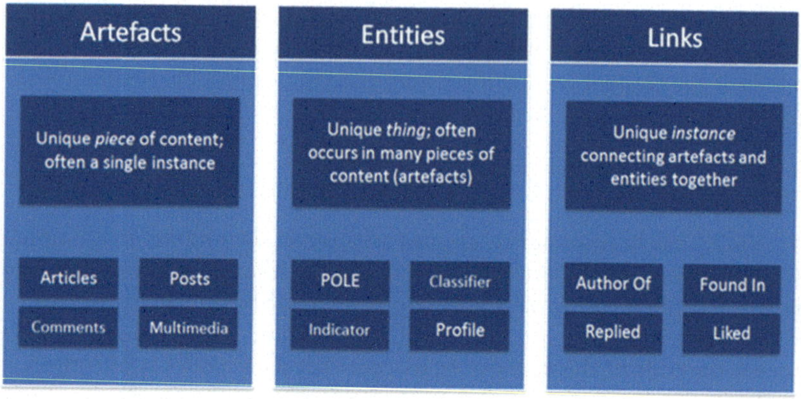

Fig. 5.1 Abstract types of content in the TENSOR data model

Artefacts represent a unique piece of content ingested by TENSOR's content acquisition functionalities. Each artefact is assigned a unique reference aligned to its source, and a content hash is created. For text, artefacts include news articles, social media posts, status updates (e.g. tweets), comments, replies or messages between individuals. As for multimedia, this includes audio, images and videos but also documents, other files and binary data.

Entities on the other hand are unique 'things' within an artefact's content or about the content (e.g. a classification or category) as well as things like social media profiles. Within a piece of content, an entity is then an attributable thing which is extractable such as a location, person, group, organisation, date, time, URLs, other social media profiles, domain names or websites. A common way to think about entities is using the POLE acronym, often referred to by law enforcement (College of Policing 2019), which standards for 'people, objects, locations and events'. In short, entities are things that link to content that are not themselves content.

Finally, between artefacts or entities are the interactions or relationships between them; these are called 'links'. Links are captured by extracting observed entities mentioned within artefacts or by observing the relationships between artefacts and entities. For example, a social media profile entity is linked to a social media post artefact with the link 'author of' if they have written the post. The extraction of links may seem trivial, but it results in a rich data set with extensive opportunities for querying data.

For example, links capture huge networks of entities based on a wide variety of relationship types such as profile mentions which can then be exploited by social network analysis to discover communities of social media profiles based around interactions, rather than simple friend/follower relationships.

Whilst the data model may be clear to those with technical background, expecting end users to become experts in understanding and appreciating abstract concepts such as artefacts, entities and links could introduce a steep learning curve, prove a barrier to long-term adoption or simply introduce unnecessary confusion during analysis. In the TENSOR intelligence dashboard, the main graphical user interface of the system, these abstract notions are hidden and instead are referred to through the natural language of the domain, e.g. posts, messages, pages, URLs, tags, profiles, persons and groups, each with their own specialised interface and representation to help extract the most meaningful knowledge for the user. Such naming conventions are also retained when displaying the graph to the user. The example in Fig. 5.2 shows a simplified version of the

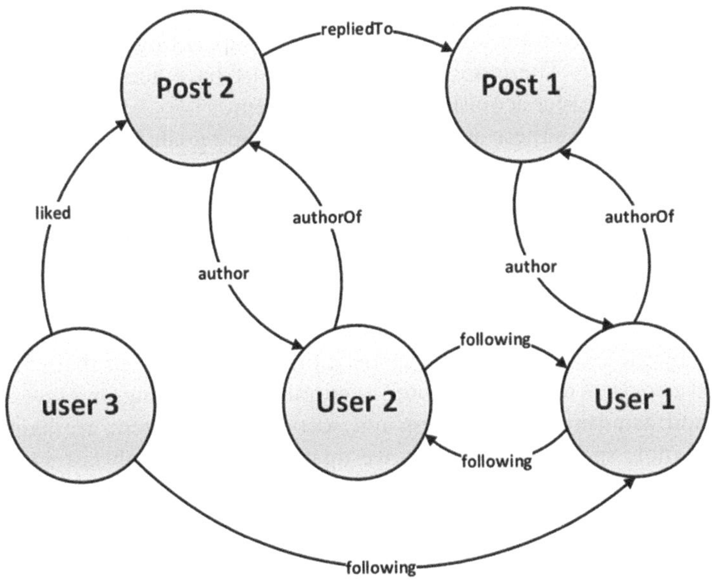

Fig. 5.2 An extract of a set of potential relationships between two posts and their authors

complex relationships between only two artefacts and three entities. In reality however, there are vastly more links extracted from even the simplest of artefacts.

5.4 Acquiring Content

Illegal and harmful content exists everywhere on the Internet, from the open web to social media platforms and forums to the Dark Web. To access such content and assess it for intelligence or evidential value may require an investigator or analyst to manually trawl through site after site, page after page, using their instinct and experience to follow potential leads. Such leads may be profiles representing known individuals or groups, content such as illegal propaganda or damaging violent images or videos. Not only is this extremely time-consuming, but it is also difficult to appreciate the extent of the content the investigator encounters and the vast number of potentially interesting and relevant connections between content, authors, profiles and other associations. These associations include followers, likes and members but also a deep network of associations within the content including an image's content, offensive and illegal references and rhetoric and who is sharing, supporting or being mentioned within it. Therefore, investigators must have access to methods which support easier acquisition of such content.

TENSOR views these as being two overarching goals for the use of the system and, depending on the action an investigator wishes to take, affects how and why content is acquired for the system. Firstly, partially reactive investigations follow a deep and narrow approach to acquiring content in that it will be focused on specific profiles representing individuals (suspects, victims) or groups (organisations). Generally, the acquisition will be interested in the individuals or groups themselves, the content being produced by these sources and any associations shared between them. These shared associates may then become a central part of the ongoing investigation and acquisition. Timeliness and scope of the content are likely to cover a more specific period; the investigation itself may only last weeks or months, whereas the content acquired (i.e. posts) may be part of a narrow window, for example, 1 week in January.

Alternatively, intelligence gathering (or proactive investigations) can be thought of as shallow and wide where larger numbers of profiles may be monitored for extended periods of time, but with less emphasis on their associates in particular. There is expected to be more emphasis on the content itself – profiles may even be generally redacted to avoid collateral collection of those on the fringes. This level of monitoring is likely to observe lots of collateral data that the investigator may never need to look at or analyse. What is more pertinent is general trends and patterns within the content over a much longer time scale, months or years.

5.5 Understanding Content

Imagine investigating a dark marketplace and finding a seller relevant to an ongoing investigation. By capturing a screenshot of their illicit product list alongside their profile name, image and a Bitcoin address, the 'content' is available for other investigators to see. This content of the screenshot makes sense to a lone investigator; they can extract the profile image, could quickly recognise if they had seen the image before, assess whether it is a generic image or perhaps specific to that user, read the Bitcoin address and recall other sellers of the same products. Even subconsciously categorising the seller based on the spectrum of their products is relatively easy, subtle and instinctive and happens without effort.

Computers struggle to match humans at this capability. What they are excellent at is processing data and making calculations, much faster than any human can. Even so, there have been many advances in the way computers can automatically interpret natural language in text and objects and scenes in multimedia. These capabilities that allow a computer system to categorise content and extract meaning from it allow investigators to take advantage of the computer's powerful abilities of organising and sorting data. Where previously an investigator would have to categorise and organise their own screenshots through post-it notes, files and index cards, with modern natural language processing (NLP) and image recognition, much of this can be automated, accelerating the scale at which information can be interpreted and an investigation can operate.

NLP and image recognition are two crucial parts of the TENSOR system as they allow content to be categorised and, therefore, organised conceptually and thematically. The first of these methods allows text to be extracted and interpreted from all types of content. The POLE acronym is an excellent starting point to understand this; it stands for 'people, objects, locations and events' and is often used by LEAs to model interacting elements of an investigation (College of Policing 2019).

Beginning with 'people', the investigator is interested in those real-life subjects who may be a suspect, victim, witnesses, bystander, foot soldier or a leader of a group. Online, such roles will be taken by authors, sharers or propagators, or even the observers, whether interacting, 'liking' or 'upvoting' content or more indirectly through their browser history or router logs. Admittedly, the last points are leaving the realms of open-source data, but nonetheless they reveal how open-source data can complement closed-source data for corroboration. People are also found within content through mentions, direct and indirect references and response. NLP can support detecting and identifying names or aliases (natural or pseudonyms), their extraction and how they are attributed across content.

'Objects' are then any tangible or intangible *thing* that can be uniquely attributed or attributed as a group of potentially unique *things*. Examples include products (cars and their models, brands and weapons), materials, media (texts, quotes, music, video), concepts (emotions, desires, beliefs, thoughts) and many other extractable *things*. 'Locations' categorise named places, meeting places, venues, parks, cities, counties and countries, whether real or fictional, globally understood (the Statue of Liberty) or only locally by a community (e.g. slang or colloquialisms). Extraction of temporal information helps with extracting 'events' and allows the system to learn about things that have happened or may in the future; it is the foundation for plotting the content and *things* within it on a timeline. Particularly relevant to TENSOR and terrorist content on the Dark Web is the detection of events and their associated timings which could indicate a forthcoming attack.

A key feature of terrorism-related content is that it is often talked about and shared in languages other than English. TENSOR currently supports translation from Arabic, French and Turkish to English directly within the system, whilst Spanish and German are supported as 'core' languages and do not require translating to be processed. The translation components make use of a number of core frameworks and services including the statistical machine translation offering Moses (Koehn et al. 2007) and a more modern neural machine translation service ModernMT (2019).

Recognising media, particularly images, uses a similar approach to extraction and attribution of textual content but is instead based on the content within the image or the image as a whole. Building on widely available image recognition models, as with NLP, TENSOR is able to exploit this vast knowledge base to categorise images into many potential groups. These models have been trained using machine learning techniques by attributing existing images, e.g. pictures of hands, landscapes, people, vehicles and all manner of other *things* in the world, to the concepts they represent. For example, there are thousands of pictures containing a hand that have been labelled with the concept 'hand'. The machine learning algorithm has used this information to produce a model that, given a new unprocessed image, is able to detect whether or not there *may* be a hand in it. The same model has also been trained with images containing thousands of other concepts, including many that could help the Dark Web investigator.

Using content acquired via TENSOR from the surface web, Dark Web and some social media platforms, the extraction of concepts and themes from text and images provides the entities and links that can be linked to the body of content. It is these entities and links that provide the data needed by many of TENSOR's analytical capabilities. The richness of these groups and connections between data make the TENSOR intelligence dashboard more powerful.

To situate these extraction capabilities, take *Alice* who is operating an illicit store front, focused on selling antique firearms favoured by gangs, on a number of dark markets. She advertises her products for short windows of time and particularly during weekends believing this will keep her off law enforcement's radar. An investigator, concerned with the illegal firearms trade, wants to begin a proactive investigation into firearms sellers on hidden dark markets. This investigator does not have any specific sellers in mind but manually scopes out the URLs of the sites they want to target. Using TENSOR, they enter the relevant URLs and begin crawls to monitor products being posted scheduling re-crawls at an hourly basis. Returning to the system after several days, they can search for images identified as containing guns or ammunition as well as textual mentions of these same concepts, given they have been extracted as entities using TENSOR's capabilities. Navigating the acquired and processed content enables quick identification of pseudonymised profiles on the marketplaces which then may warrant further investigation based on how prolific they are or their connectedness. Furthermore, all mentioned profiles are stored alongside other extracted information such as Bitcoin wallets, specific terminology or keywords or products which may be vital later in the investigation.

5.6 Analysing Content

Obtaining significant amounts of content has become easier with automated content acquisition capabilities; however, this has led to a major growth in the challenges of handling that data. Historically, systems relied on manually trawling, categorising and indexing all investigation material, but the extent of digital content available through both offline and online sources make analysing such information without automated means near impossible.

Fortunately, automated methods to extract and categorise content are becoming increasingly accurate, but this still leaves the problem of how to find the proverbial needles in the haystack? These needles may be specific pieces of content that are critical to an investigation, perhaps a hint to a potential attack, but they may well be hidden in the complex web of interactions. In TENSOR, the investigator can call on advanced automated analyses which can detect textual and visual patterns, group and organise content, find paths between entities and their content and uncover organisational structures.

5.7 Beyond TENSOR and Lessons Learned

TENSOR as a system does not stand alone, and many other systems with similar and complementary capabilities exist. TENSOR is not supposed to be a silver bullet designed to replace all existing systems. Therefore, it is crucial that TENSOR provides ways for investigators to provide ways of seeding and enriching its initial content with other sources they have access to and, in common formats such as CSV (comma-separated value), always allowing the system's potential to be exploited by law enforcement.

TENSOR must also 'close the loop' for investigators by supporting and exploiting the enrichment and analysis performed within the system and allowing investigations to continue forwards beyond TENSOR. This is imperative when considering how connections in the discovered content can lead to new relevant content and also new potential leads in the form of profiles, events and analyses. At the raw level, content can simply be exported, again in the common CSV format, allowing further exploration in tools the investigator is familiar with such as Microsoft Excel or using IBM's I2 Analyst's Notebook to further develop elements of the investigation to present in court, for example. On top of this, many of TENSOR's visualisations or analytical outputs may be captured or exported to support these purposes including the investigators notes, the commonalities and visualisations.

Connecting the inputs and outputs of TENSOR with those resources and tools available to investigators at the beginning and the end of an investigation or intelligence gathering exercise demonstrates just how TENSOR can deliver value and save time.

Throughout the development of TENSOR, opportunities for new knowledge and insights have been sought not only from a technical perspective but also directly from the terrorism domain and across the legal, ethics, data protection and other areas. The emphasis here will be more on the technical side, but these are by no means the only important takeaways for this ambitious and challenging project.

First and foremost, an area key of key importance has been gaining a deeper understanding of the challenges facing any kind of automated Deep Web acquisition. The generally accepted model, including the one used in this book, is that of an iceberg with three layers: the surface web, the Deep Web and the Dark Web. Whilst this is a reliable model for thinking about the layers of the web, it is also useful to think about the Dark Web as a more logical layer which can have its own surface and Deep Web within it. These surface and Deep Webs may also occur across many darknets as well. Darknets being the additional layers of the web that are hidden across different protocols, such as Tor's hidden services, I2P or Freenet. The first lesson here is on the assumption that because a human being may find it relatively straightforward to access these services in the Deep Web, this is not the same for a system such as TENSOR. Captchas are a great example; these are the fuzzy texts requesting the user to enter them correctly to check that they are a human. People generally do not struggle with them, computers on the other hand do – the actual point of having a Captcha. These Captchas only exist because they are effective, and many Deep Web sites and services use them for this reason. Secondly, accessing these services as a human being has a very particular usage profile that is difficult for a computer to replicate. If the acquisition task was to be automated in the same way a human uses the service and via a web browser, it is likely that the pattern of usage would differ in very detectable ways from a human user.

As a result of this challenge, the more common route when it comes to Deep Web sources such as social media platforms is to utilise their existing application programming interfaces (APIs). This is also based on the assumption that these social media platforms all have APIs, when in fact the majority of them do not. Normally, they are only provided by larger platforms, and these are quickly becoming more difficult to use or being ruled out based on their terms of service. For example, many services now declare that law enforcement is not to use the APIs or that they cannot be used for any activities resembling surveillance, i.e. the continued monitoring of an account. Additionally, APIs are also a moving target as platforms continue to develop. APIs also create a development overhead as the more services are integrated the more integrations need to be maintained.

This leads onto the next emerging challenge with the Deep Web: the scale of growth in the app-only market and other sources that are challenging to access. Services that operate solely through mobile apps are plentiful and continue to grow in popularity; these include popular services such as WhatsApp and Snapchat, and although this is changing – for example, WhatsApp can be accessed through a web browser – access remains limited. Furthermore, with each service operating in a different way, it is difficult for investigators to keep track of where illegal activity may be being conducted regardless of whether they can access such content. Another known issue of access to a Deep Web source is that of in-game communication systems that are increasingly popular in modern advanced gaming but are extremely difficult to extract any information from automatically. Ultimately, better cooperation between service providers and law enforcement, built on a sense of trust from both sides, will allow capabilities such as TENSOR to be utilised safely and fairly by law enforcement for the purposes of protecting the public.

In software development terms, all systems, especially those which have been co-developed by academia and industry, have to find a balance between performance that can be achieved in a laboratory environment and state-of-the-art implementation of algorithms and the requirement for future operation deployment on systems which may have limited hardware capacity. However, as Goble (2014) notes, developing better research software has a wide range of benefits to both the researchers and science as a whole. TENSOR has also experienced that what works in the lab does not always experience the same high-level performance when tested by real users or that particular challenges may arise when an investigator needs to re-run an analysis but also maintain the chain of evidence for their existing analyses. Additionally, even only when simulating operational use does it become clear the true extent of the volume of data law enforcement is encountering when they need to investigate openly accessible data sources. Nonetheless, it is also when such volumes of data confront the developers does the requirement to pull from advanced research techniques become even more apparent.

Furthermore, a key lesson from the project has been to realise, often, how seemingly simple functionalities that are neither difficult to imagine nor implement are missing from existing analytical software used by law enforcement. Occasionally, it is these solutions that provide the most value to users, such as the commonalities view in TENSOR, and can actually increase adoption and uptake of the system where the advanced functionalities can come into play.

5.8 Concluding Remarks

This chapter has set down the power of TENSOR as an Internet intelligence and investigation tool. It has demonstrated the layered capabilities of TENSOR that is able to collectively deliver a coherent and widely applicable data model that generically supports a wide range of web, social media, forum and marketplace sites across the surface, Deep and Dark Web.

On the top layer, TENSOR supports a single central intelligence dashboard through which all content can be accessed, whilst hiding the complexity of the processing underneath and focuses on delivering value and valuable insights. In the future, some interfaces will support direct access to the specific functionalities of certain components of the system – for example, an investigator may have an image they need to upload to check for evidence of tampering but does not need to acquire any additional data around that image at that point in time.

The development of TENSOR has also demonstrated the ways in which even software still under development can enhance investigatory approaches, especially if the overall system is not intended to be a silver bullet. Whilst some modules offer an experimental approach, taking advantage of the latest research and algorithms, other components, such as the commonalities interface, can deliver immediate results. Other aspects are continually improving, specifically; advances in NLP, multimedia extraction and computer vision are almost continuous and can often be easily integrated.

As with any big data system, you cannot avoid capturing data that ultimately turns out to be noise. This is particularly true for the processing steps after content acquisition where false positives of concepts may be extracted from text or objects from images. However, corroboration between data, using techniques such as FCA, can help reduce these false positives by linking content based on similarities and ignoring or diminishing the importance of the outliers.

Ultimately, the goal of the TENSOR system is to ensure that law enforcement can access and analyse complex information from the surface, Deep and Dark Web, whilst supporting advanced analytical capabilities that deliver real operational value to the user. The expert decision-making is left to the human in the loop, whilst the monotonous and data-heavy processing is performed by the system. Symbiotically, TENSOR is able to deliver extensive intelligence generated by users without the need for extensive technical knowledge that restricts adoption.

References

B. Akhgar, P. Bertrand, C. Chananouli, T. Day, H. Gibson, D. Kavallieros, I. Kompastsiaris, E. Kyriakou, G. Leventakis, E. Lissaris, S. Mille, T. Tsikrika, S. Vrochidis, U. Williamson, TENSOR: Retrieval and analysis of heterogeneous online content for terrorist activity recognition. Proceedings Estonian Academy of Security Sciences, 16: From Research to Security Union **16**, 33–82 (2017)

J. Belur, S. Johnson, Is crime analysis at the heart of policing practice? A case study. Polic. Soc. **28**(7), 768–786 (2018)

College of Policing, *Collection and recording. College of Policing Authorised Professional Practice* (2019, August), Retrieved from https://www.app.college.police.uk/app-content/information-management/management-of-police-information/collection-and-recording/

Deloitte, *The Digital Policing Journey: From Concept to Reality – Realising the Benefits of Transformative Technology* (2015), Retrieved from https://www2.deloitte.com/content/dam/Deloitte/uk/Documents/public-sector/deloitte-uk-ps-digital-police-force.pdf

B. Feldman, *It Is Incredibly Easy to Fake a Screenshot. Here's How.* Intelligencer (2015, December 11), Retrieved from http://nymag.com/intelligencer/2015/11/how-to-fake-a-screenshot.html

G. Giova, Improving chain of custody in forensic investigation of electronic digital systems. Int. J. Comput. Sci. Netw. Secur. **11**(1), 1–9 (2011)

C. Goble, Better software, better research. IEEE Internet Comput. **18**(5), 4–8 (2014)

P. Koehn, H. Hoang, A. Birch, C. Callison-Burch, M. Federico, N. Bertoldi, B. Cowan, W. Shen, C. Moran, R. Zens, C. Dyer, O. Bojar, A. Constantin, E. Herbst, Moses: Open source toolkit for statistical machine translation, in *Proceedings of the 45th Annual Meeting of the Association for Computational*

Linguistics Companion Volume Proceedings of the Demo and Poster Sessions (2007), pp. 177–180

M. Langheinrich, Privacy by design—Principles of privacy-aware ubiquitous systems, in *Proceedings of the International Conference on Ubiquitous Computing*, (Springer, Berlin, Heidelberg, 2001), pp. 273–291

ModernMT (2019), Retrieved from https://www.modernmt.com/

PART II

Legal and Ethical Considerations

CHAPTER 6

Ethical and Societal Issues of Automated Dark Web Investigation: Part 1

Marco Gercke

6.1 Introduction

The purpose of these five chapters is not to determine whether an LEA in one of the EU member states will be authorised to carry out investigations using a specific automated tool or whether evidence collected within the application of the tool will be admissible in court. These important questions will require a case by case evaluation by the institution that plans to utilise the specific tool.

The main focus of these five chapters is to identify potential challenges for the application of a sophisticated automated tool. Such challenges could require the limitation of functions of the tool already in the design process. Any development of such a tool has to be committed to "ethical and legal compliance by design".

M. Gercke (✉)
Cybercrime Research Institute, Cologne, Germany
e-mail: gercke@cybercrime.de

B. Akhgar et al. (eds.), *Dark Web Investigation*,
Security Informatics and Law Enforcement,
https://doi.org/10.1007/978-3-030-55343-2_6

6.2 Methodology and Approach

According to Europol's EU Terrorism Situation and Trend Report (TE-SAT) 2017,[1] terrorists use public communication to attract potential recruits, procure material and financial support and intimidate opponents. In 2016 terrorist groups continued to use online services for communication in targeted and diverse ways: terrorist propaganda was spread primarily through social media platforms and file-sharing sites, while encryption and anonymisation technologies (e.g. the encrypted social media platform Telegram) maintained the anonymity of terrorists.[2] Terrorist groups increased their use of other encrypted social media platforms, most notably Telegram, and their supporters have used it as a starting point for moving to platforms with a stronger impact, in particular Twitter and, to a lesser extent, Facebook.[3] In late 2016, Facebook, Microsoft, Twitter and Google (YouTube) created a joint database to quickly identify and remove images and videos promoting terrorism from their platforms.[4] The noticeable disruptive effect of this measure encourages any efforts to develop a helpful automated tool for LEAs.

To counteract this increase of terrorist activity online, the vision for a sophisticated tool is to automate key preparatory activities of LEAs currently binding a lot of time and resources. It may be envisioned that terrorism prevention units search the surface web and the dark web for suspicious names, places, websites or activities as selectors with the help of an automated tool. The search result presented by an automated tool then not only reveals in what ways the searched (id)entity appears connected to terrorist material and/or activities but also is capable of tracing additional suspicious activities as well as (id)entities.

Any use of a tool by LEAs takes place within the framework of society at large. As such, the society (at large and within each European member state) has an impact on how this tool may be used. The core aim of Chaps. 6, 7, 8, 9 and 10 is to identify and to analyse any real and potential legal,

---

[1] Europol (2017a).

[2] Europol (2017b).

[3] Europol (2017b).

[4] When one company identifies and removes such a piece of content, the others will be able to use a unique digital fingerprint (hash) to identify and remove the same piece of content from their own network in accordance with their own policies. See "Partnering to Help Curb Spread of Online Terrorist Content", Facebook Newsroom, 5 December 2016.

political, ethical and societal implications for the use of a sophisticated automated tool.

For the guidance provided in this book to serve with lasting effect, it seemed appropriate to separate the analysis of the legal framework from the analysis of the political, ethical and societal framework for the use of a sophisticated automated tool. It follows that the two main questions considered are:

1. What are ethical implications for the use of an automated tool by LEAs?
   The analysis of this question is provided in Chaps. 7 and 8 based on the methodology outlined in Sect. 6.3.
2. What are the global, regional and national legal frameworks for using a sophisticated automated tool and how do they interact?
   The analysis of this question is provided in Chaps. 9 and 10 based on the methodology outlined in Sect. 6.4.

6.3 Identifying Political, Ethical and Societal Issues Concerning an Automated Tool

In an effort to identify the key political, ethical and societal issues that had to be covered in Chaps. 7 and 8, a traditional PESTLE or STEP approach was carried out. Based on this traditional method, a sophisticated tool automating significantly the efforts for the prevention of terrorism involves three major moral objections: *first*, the quality of evidence produced by the tool; *second*, the risk of potential bias; and *third*, the danger of unfair effects based on the tool's results.

For the purposes of focus control, a red teaming exercise[5] was then carried out to identify possible gaps of the traditional and rather general PESTLE or STEP approach. The red teaming exercise revealed four additional aspects which are typically relevant when an automated tool is used by LEAs.

First, it appears ethically questionable whether an automated tool's degree of surveillance is in the interest of the people whom state authorities are established to protect. Any form of surveillance not only questions the value of private life but also carries the risk of inhibiting the expression of the will of the people and the exercise of democratic rights such as the freedom to oppose a government.

[5] For an explanation of "red teaming", see Sect. 6.4.

Second, the effectiveness of using an automated tool appears in doubt because it will intrude deeply into the lives of individuals who are not involved in terrorism and against whom there is either no or non-compelling evidence of wrongdoing while neither the probability of terrorist harm is high nor its source established.

Third, it seems ethically challenging to identify appropriate search terms (selectors) for an automated tool to identify suitable evidence because such selectors should be adequately specific, reasonable, evidence-based and non-discriminatory.

Fourth, there is the risk of chilling effects on individuals whom the use of the automated tool might disincline from engaging in otherwise perfectly legitimate online activities. Such chilling effects emanate from the fear of unjustified stigmatisation and are at odds with democratic values and practice.

Taking all political, ethical and societal considerations together, the key issues concerning an automated tool can be structured into two sections. First, Chap. 7 focusses on the examination of the:

1. Quality of evidence produced (see Sect. 7.3 in Chap. 7)
2. Danger of unfair effects based on the tool's results (see Sect. 7.4 in Chap. 7). Second, Chap. 8 focusses on the investigation of the:
3. Compatibility with the democratic interest of the people (see Sect. 8.1 in Chap. 8).
4. The effectiveness of the tool (see Sect. 8.2 in Chap. 8).
5. The identification of appropriate selectors (see Sect. 8.3 in Chap. 8) and,
6. The risk of chilling effects (see Sect. 8.8 in Chap. 8).

6.4 Identifying Key Legal Implications and the Relevant Legal Framework

For the purposes of focus control, a red teaming exercise was carried out to identify possible topics that need to be included in a review.

Red teaming or alternative analysis is a specific method used to review plans, strategies and hypotheses.[6] Two teams are formed, a red team and a blue team.[7] The red team assumes the role of the attacker, while the blue

[6] See: Herman et al. (2009), Sabin (2012), Fryer-Biggs (2012), Lauder (2009), Longbine (2008), Wood and Duggan (2002).

[7] See Wood and Duggan (2002).

team focusses on defence.[8] This method has been successfully employed by the military for decades[9] and has also been applied in civil activities for a number of years.[10] It is explicitly not restricted to acting out physical attacks. The methodology can also be used to investigate theoretical issues from different angles and with varying emphases – reaching as far as intangible constructs such as a legislative draft.[11] Red teaming can be particularly useful when developing cybersecurity strategies, since the attack situation reflects the real threat situation. However, strategies are mostly developed from the defence angle. A change or expansion of perspective enables a company's own strategies to be examined more critically. Red teaming is not limited to military context, but it can even be utilised in the process of drafting legislation.[12]

The red teaming exercise revealed the following 11 legal issues that are included in the analysis provided in Chap. 9:

1. LEAs need to be aware that despite harmonisation approaches the legislation legal framework criminalising terrorist activities differs from country to country.
2. For any LEA the use of any automated tool has to be based on a legitimate legal basis which includes the use of an automated tool independently of a specific case.
3. If different LEAs are involved in antiterrorism operations, the mandate of each LEA to use the same automated tool needs to be accessed individually, and certain functions might need to be restricted for some LEAs.
4. Using a tool that is able to automate searches and collect large volumes of data may not be covered by provisions authorising the search and collection of data in a manual fashion.
5. When implementing undercover operation capabilities, it is important to design it in accordance with national legal frameworks and practices.
6. The collection of large volumes of data may include personal data and consequently raise questions with regard to data protection.

[8] See Meija (2008).
[9] See Lauder (2009), and Longbine (2008).
[10] See Lauder (2009).
[11] See Gercke (2014).
[12] See Gercke (2014).

7. Cross-border collection of information raises questions related to national sovereignty, and exchanging information cross border might fall under formal agreements related to international cooperation in criminal matters.
8. During the collection of terrorist content, illegal content might be collected. The possession of such material could lead to criminal investigations.
9. Collecting and storing media that was published online could violate copyright laws.
10. Collecting information from social media platforms could violate licensing or end user agreements.
11. If the data collected by an automated tool should be used as evidence in court, it will be important to ensure that the rules and regulations with regard to the collection of evidence are observed when designing the tool.

Having identified these 11 key areas of legal implications of the use of an automated tool by LEAs, the research for Chaps. 9 and 10 was structured into 3 steps supplementing and building on each other.

As *step 1*, research was carried out to create and present a comprehensive inventory of relevant publications. This has led to a thorough overview over already existing scientific explorations and examinations of legal implications for each topic.

As *step 2* and drawing from the inventory established in step 1, the general aspects are presented in Chap. 9 which are specifically relevant for each of the 11 topics.

As *step 3*, the general aspects presented in step 2 are contrasted in Chap. 10 with the legal framework first at global (United Nations) and then at regional level (Council of Europe and European Union).

6.5 Developments in Terrorist Use of the Internet and Investigations: How Terrorist Use of the Internet Changed

Today the Internet is widely considered a means of communication with great potential for connecting people. Since the 1990s the total number of people using the Internet has increased year by year.[13] Especially the rise of social media changed the way people are communicating today.

[13] See for more details: ITU, ICT Facts & Figures 2015, 2015.

At the same time, the same technology moved into the focus of terrorist organisations and criminals.[14] It is widely recognised that terrorists are using Internet services for their purposes.[15] Research shows that the Internet is used by terrorist organisations at large. In the 1990s[16] the discussion about the use of the network by terrorist organisations focussed on network-based attacks against critical infrastructures and the use of information technology in armed conflicts. This view began to change after the 9/11 attacks.[17] Although not directly cyber-related, it was reported[18] that the 9/11 offenders used the Internet during the preparatory stage of the attacks.[19] [20] With the new debate about terrorist use of the Internet, various different ways in which terrorist organisations use the Internet were identified.[21] The March 2009 Report of the CTITF Working Group on Countering the Use of the Internet for Terrorist Purposes already listed fundraising, training, recruitment, secret communication, data mining, propaganda and radicalisation.[22] More recently founded terrorist groups such as ISIS utilise the Internet as well as the dark web and modern communication technology even more intensively.[23]

It is therefore a logic consequence that terrorist use of the Internet and dark web is moving into the focus of LEAs.

[14] Report of the Working Group on Countering the Use of the Internet for Terrorist Purposes, 2009, page 3.

[15] See, for example, CTITF (2011), CTITF (2009), UNODC (2012).

[16] Gercke (2007).

[17] See: Lewis (n.d.-a, n.d.-b), Gercke (2007), Sieber and Brunst (2007), Denning (n.d.), Embar-Seddon (2002); United States Department of State, "Pattern of Global Terrorism, 2000", in: Prados, America Confronts Terrorism, 2002, 111 et seq.; Lake (2000), Gordon (n.d.), US-National Research Council (2003), and OSCE/ODIHR (2007).

[18] See: Rötzer (2001).

[19] The text of the final message was reported to be: "The semester begins in three more weeks. We've obtained 19 confirmations for studies in the faculty of law, the faculty of urban planning, the faculty of fine arts, and the faculty of engineering."The name of the faculties was apparently the code for different targets. For more detail see Weimann (2005), Thomas (2003), and Zeller (2004).

[20] CNN (2004).

[21] For an overview see: Sieber and Brunst (2007), Gercke (2007).

[22] Report of the Working Group on Countering the Use of the Internet for Terrorist Purposes, 2009, page 5–8.

[23] See, for example, Siboni et al. (2015), Hoffman and Schweitzer (2015).

6.6 New Developments: Artificial Intelligence and Machine Learning

When analysing the technical developments in more detail several significant trends can be observed. Technical developments such as artificial intelligence and machine learning are not only making significant progress – they have the potential to fundamentally change the role of technology. Already today computer systems carry out various tasks faster and more accurate than human beings. And this is not anymore limited to typical processes such as calculation and interaction with large data volumes.

The success of AI-based systems in beating humans by applying deep learning and deep reinforcement learning underlines in a very illustrative way the progress of this field.[24] The fact that AI systems are able to quickly pick up the rules of the game without being taught in advance attracted a lot of attention even outside the scientific community.[25] The underlying developments go beyond video games and started significantly earlier.[26] Other visible signs of the integration of AI are, for example, successful tests with self-driving cars.[27]

It is, therefore, not surprising that an evaluation on how these technologies can be utilised by LEAs can be beneficial to enhance investigation.[28] This discussion is going on for some years – often with a focus on data mining.[29] An example is COPLINK, an integrated information management system developed by the University of Arizona. It allows more advanced searches in database and the exchange of information among LEAs.[30]

It is currently difficult to say if the progress of automation and the increasingly rapid advances in and application of artificial intelligence (AI) present more of a challenge or an opportunity for law enforcement.

But recent technical developments in the field of linguistic and speech interaction allow going beyond this. In 2014 a machine for the first time

[24] See, for example, Mnih et al. (2015).
[25] See, for example, McMillan (2015).
[26] See Mnih et al. (2013).
[27] See, for example, KMPG (n.d.).
[28] See, for example, Alzou'bi et al. (2014).
[29] Holmes et al. (2007a).
[30] See Schroeder (2001), and Holmes (2007b).

passed the Turing test.[31] The test is based on a communication between an evaluator and two partners.[32] The evaluator is aware that one of the two partners in conversation is a machine. All three participants are separated from one another. The conversation is limited to text only. If the evaluator is unable to tell the machine from the human, the machine is said to have passed the Turing test.

The ability of machines to hold text conversations and, for example, participate in online discussion forums, leave comments related to publications and engage in chats with individuals offers new possibilities for investigators. A LEA-specific software could take over investigations – e.g. monitoring of radicalisation in online forums – that currently require manual work of investigators as mere passive observation appears suspicious and makes it rather easy for members of a discussion forum to identify participants that solely intend to observe or record conversations.

Artificial intelligence and big data analysis have the potential to provide significant input to the work of law enforcement[33] – if it is possible to overcome legitimate concerns related to data protection and fundamental human rights.

6.7 Use of Technology by LEAs and Legal Considerations

For the last 30 years, law enforcement agencies (LEAs) were among those professionals that benefitted from technical developments. One example is the automated search for known child pornography images. While in the past forensic experts or police officers needed to manually search for illegal content on a suspect's computer system or storage device, such processes were automated shortly after the underlying technology was available. This process has continued ever since. Technology aims to provide LEAs with tailored tools to support their work in preventing and investigating terrorist activities.

The main focus of developing a sophisticated automated tool would have to be on identifying the end user demands and matching them with the technical solutions and methods that are available to meet the demands. Within such focus technology is the limiting factor. Or in other words,

[31] Warwick and Shah (2015).

[32] Turing (1950).

[33] See, for example, Alzou'bi et al. (2014).

solutions can be developed if the end users express demand and technical solutions are available.

While in the past technical limitations were usually the main obstacle in the development of more and more efficient solutions to support LEAs, the incredible technical developments having taken place in recent years required to add certain criteria to the drafting and development process. Ethical questions are gaining an increasing role. They are addressed in Chaps. 7 and 8.

But ethical considerations are not the only ones of relevance. Over the last years, a complex legal and regulatory framework (see Chaps. 9 and 10) has been developed that addresses various aspects of data interaction. One example is data protection. Lawmakers and international organisations responded to an increasing interaction with personal data by creating a legal framework to protect such data.

When it comes to LEAs, legal frameworks are in general of great importance as LEAs cannot simply apply any technology that is available on the market. Especially in continental European states, it is necessary that statutory law provides a legal basis for the investigation in question.

References

Alzou'bi, Alshiibly, AlMa'aitah, Artificial intelligence in law enforcement, a review, Int. J. Adv. Inform. Technol **4**(4), 1 (2014) et seq

CNN, News, 4 Aug 2004. Available at: http://www.cnn.com/2004/US/08/03/terror.threat/index.html

CTITF, *Report of the Working Group on Countering the Use of the Internet for Terrorist Purposes* (2009)

CTITF, *Countering the Use of the Internet for Terrorist Purposes – Legal and Technical Aspects* (2011)

D. E. Denning, Activism, hacktivism, and cyberterrorism: The Internet as a tool for influencing foreign policy, in *Networks & Netwars: The Future of Terror, Crime, and Militancy*, ed. by J. Arquilla, D. Ronfeldt, p. 239 et seq. Available at: http://www.rand.org/pubs/monograph_reports/MR1382/MR1382.ch8.pdf (n.d.)

A. Embar-Seddon, Cyberterrorism, are we under siege? Am. Behav. Sci **45**, 1033 (2002) et seq

Europol, *EU Terrorism Situation and Trend Report (TE-SAT)* (2017a). Available at: https://www.europol.europa.eu/activities-services/main-reports/eu-terrorism-situation-and-trend-report-te-sat-2017

Europol, *EU Terrorism Situation and Trend Report (TE-SAT)* (2017b), p. 29

Z. Fryer-Biggs, *Building Better Cyber Red Teams*, defensenews.com. 14 June 2012
M. Gercke, *Cyberterrorism, How Terrorists Use the Internet, Computer und Recht* (2007), p. 62 et. seq
M. Gercke, *"Red Teaming" Ansätze zur Effektivierung von Gesetzgebungsprozessen?* Die Übertragbarkeit einer klassischen, militärischen Methodik auf Gesetzgebungsprozesse im IT-Bereich, CR (2014), p. 344 et seq
S. Gordon, *Cyberterrorism*. Available at: http://www.symantec.com/avcenter/reference/cyberterrorism.pdf (n.d.)
M. Herman, M. Frost, R. Kurz, *Wargaming for Leaders* (2009)
A. Hoffman, Y. Schweitzer, *Cyber Jihad in the Service of the Islamic State (ISIS)* Strategic Assessment, by The Institute for National Security Studies (INSS), Tel Aviv University Strategic Assessment, by The Institute for National Security Studies (INSS), Tel Aviv University **18**(1), 71 et. seq (2015)
M. C. Holmes, D. D. Comstock-Davidson, R. L. Hayen, Data mining and expert systems in law enforcement agencies, Issues. Inform. Syst **VIII**(2), 329 (2007a) et seq
M. C. Holmes, D. D. Comstock-Davidson, R. L. Hayen, Data mining and expert systems in law enforcement agencies, Issues. Inform. Syst **VIII**(2), 330 (2007b) et seq
KMPG, *Self-Driving cars: The Next Revolution*. Available at: https://www.kpmg.com/US/en/IssuesAndInsights/ArticlesPublications/Documents/self-driving-cars-next-revolution.pdf (n.d.)
Lake, *6 Nightmares* (2000), p. 33 et seq
M. Lauder, Red Dawn: The emergence of a red teaming capability in the Canadian Forces. Can. Army. J **12**(2), 25–36 (2009)
J. A. Lewis, *The Internet and Terrorism*. Available at: http://www.csis.org/media/csis/pubs/050401_internetandterrorism.pdf (n.d.-a)
J. A. Lewis, *Cyber-Terrorism and Cybersecurity*. http://www.csis.org/media/csis/pubs/020106_cyberterror_cybersecurity.pdf (n.d.-b)
D. F. Longbine, *Red Teaming: Past and Present* (2008)
R. McMillan, Google's AI is now smart enough to play Atari like the Pros, Wired Magazine (2015)
R. Meija, *Red Team Versus Blue Team – How to Run an effective Simulation*, CSO 25 Mar 2008
V. Mnih, K. Kavukcuoglu, D. Silver, A. Graves, I. Antonoglou, D. Wierstra, M. Riedmiller, *Playing Atari with Deep Reinfocement Learning*, NIPS 2013 Workshop paper (2013)
V. Mnih, K. Kavukcuoglu, D. Silver, A. A. Rusu, J. Veness, M. G. Bellemare, A. Graves, M. Riedmiller, A. K. Fidjeland, G. Ostrovski, S. Petersen, C. Beattie, A. Sadik, I. Antonoglou, H. King, D. Kumaran, D. Wierstra, S. Legg, D. Hassabis, Human-level control through deep reinforcement learning, Nature, **518**, 529 (2015) et seq

OSCE/ODIHR Comments on legislative treatment of "cyberterror" in domestic law of individual states (2007). Available at: http://www.legislationline.org/upload/lawreviews/93/60/7b15d8093cbebb505ecc3b4ef976.pdf

Rötzer, *Telepolis News*, 4 Nov 2001. Available at: http://www.heise.de/tp/r4/artikel/9/9717/1.html

P. Sabin, *Simulating War* (2012)

J. Schroeder, *Coplink: Database Integration and Access for a Law Enforcement Intranet*, 2001

G. Siboni, D. Cohen, T. Koren, The Islamic State's Strategy in Cyberspace. Mil. Strateg. Aff 7(1), 127–144 et. seq (2015)

U. Sieber, P. Brunst, Cyberterrorism and other use of the Internet for terrorist purposes : Threat analysis and evaluation of international conventions. Codexter (Council of Europe, Strasbourg), 03 R, 4–77. (2007)

Thomas, Al Qaeda and the Internet: The danger of "cyberplanning" (2003). Available at: http://findarticles.com/p/articles/mi_m0IBR/is_1_33/ai_99233031/pg_6

A. M. Turing, Computing machinery and intelligence, Mind.New. Series. **59**(236), 433 (1950) et seq

UNODC, *The Use of the Internet for Terrorist Purposes* (2012)

US-National Research Council, *Information Technology for Counterterrorism: Immediate Actions and Future Possibilities* (2003), p. 11 et seq

K. Warwick, H. Shah, Can machines think? A report on Turing test experiments at the Royal Society, J. Exp. Theor. Artif. Intell, 1 (2015) et seq

G. Weimann, How modern terrorism uses the internet. J. Int. Secur. Aff (8), (2005)

B. J. Wood, R. A. Duggan, Red teaming of advanced information assurance concepts, in *DARPA Information Survivability Conference and Exposition. DISCEX 00 Proceedings*, vol 2 (2002), S. 112ff

T. Zeller, *On the Open Internet, a Web of Dark Alleys*, The New York Times, 20 Dec 2004.Availableat:http://www.nytimes.com/2004/12/20/technology/20covert.html?pagewanted=print&position=;

CHAPTER 7

Ethical and Societal Issues of Automated Dark Web Investigation: Part 2

Ulrich Gasper

7.1 Political, Ethical and Societal Issues Concerning an Automated Tool

The use of the envisioned sophisticated tool may be intended to automate investigations by antiterrorism units which are currently carried out by humans. For that purpose, the tool should provide reliable data concerning terrorist material and activities which may serve as evidence (justifying at least further investigations if not already reliable in court) and trigger further action by a LEA.

7.2 Algorithms and Machine Learning

In order to allow and fulfil these intended functions, the automated tool has to employ algorithms. Although an "algorithm" may formally be defined as purely *mathematical* construct (Hill 2016), lay usage of the

U. Gasper (✉)
Cybercrime Research Institute, Cologne, Germany
e-mail: gasper@cybercrime.de

B. Akhgar et al. (eds.), *Dark Web Investigation*,
Security Informatics and Law Enforcement,
https://doi.org/10.1007/978-3-030-55343-2_7

term "algorithm" also includes the implementation of the mathematical construct into a technology and an application of the technology configured for a particular task.[1] A fully configured algorithm incorporates the abstract mathematical structure that has been implemented into a system for analysis of tasks. Whereas a strict wording would have to distinguish between constructs, implementations, and configurations, for the discussion of ethical issues in this deliverable generically referring to "algorithm" will suffice. Algorithms in this sense will not only be found in the configuration of the automated tool itself but also in the configuration of the external search engines like Google, Bing, DuckDuckGo, Torch, Ahmia and others which the tool is intended to take advantage of and incorporate into its operations.

Replacing a human operator of an investigation at least to a significant extent by an algorithm has the advantage that the analysis is augmented already by the scope and scale of data and rules involved. The way in which an algorithm makes sense of streams of data and determines features relevant to a given decision outperforms any human operator and involves a qualitatively different decision-making logic applied to larger inputs. The new scale of analysis and the complexity of decision-making are already ethically challenging, and this challenge is increased by the opacity of the work by employed algorithms. Traditionally, algorithms operate on decision-making rules which are defined and programmed individually "by hand" (e.g. Google's PageRank algorithm), but increasingly rely on machine learning capacities which are also referred to as "predictive analytics"[2] and "artificial intelligence"[3] because these algorithms are capable of learning (Tutt 2017). Also a sophisticated tool will ultimately be envisioned to have capacities of machine learning. Machine learning generally means that the algorithm defines the decision-making rules to handle new inputs independently of a human operator.[4] Such learning capacities grant algorithms a degree of autonomy; the impact of which ultimately remains uncertain. As a result, tasks performed by machine learning algorithms are not only difficult to predict beforehand but also

[1] See Turner and Angius (2017, p. 47).
[2] See (Siegel 2016).
[3] See (Domingos 2015).
[4] Matthias (2004, p. 179).

difficult to explain afterwards, and this uncertainty might inhibit the identification and redress of ethical challenges.[5]

Against this background, this chapter suggests three major ethical challenges for the use of a sophisticated tool:

1. The quality of evidence produced by an automated tool (see Sect. 7.3 - 7-5)
2. The risk of potential bias (see Sect. 7.6)
3. The (un)fairness of the actions driven by an automated tool (see Sect. 7.7 - 7.12)

7.3 Quality of Evidence

The first major ethical challenge posed by an automated tool which is influenced by decision-making algorithms concerns the quality of evidence produced by the algorithm. It seems appropriate to divide this challenge in the following three components[6]:

- The (in)conclusiveness of evidence (see Sect. 7.4)
- (In)scrutability of evidence (see Sect. 7.5)
- The risk of potential bias (see Sect. 7.6)

7.4 (In)Conclusiveness: Correlation or Causality

Algorithmic decision-making and data mining rely on inductive knowledge and correlations identified within the data examined. The evidence produced by an algorithm does not establish any causality. The necessary search for causal links is complicated by the phenomenon that correlations based on a sufficient volume of data could increasingly be seen as sufficiently credible to direct actions without first establishing causality.[7] Acting upon mere correlations may ethically be legitimate but requires a higher threshold of evidence to justify actions with ethical impact. The risk is that algorithmic categories signal certainty, discourage alternative explorations

[5] Mittelstadt et al. (2016, p. 3).

[6] Mittelstadt et al. (2016, p. 4), referring to the quality of evidence as "inconclusive", "inscrutable" and "misguided".

[7] Hildebrandt (2011, pp. 378–380).

and create a coherence among disparate objects.[8] This leads to the danger of having individuals described via too simplified models (Barocas n.d.). This risk, as well as this danger, appears to be manageable by the fact that the search results of the automated tool will be evaluated by an officer of an LEA especially if such officer has been trained to take both the risk of false certainty and the danger of (over)simplification into account.

7.5 (In)Scrutability of Algorithm's Functionality and Rationale

The scrutability of evidence presents an essential ethical concern and addresses the transparency and opacity of an algorithm. The primary components of transparency are accessibility and comprehensibility of information, but information about the functionality of algorithms is often poorly accessible. Proprietary algorithms are kept secret either for the sake of competitive advantage[9] or of national security.[10] The transparency of an algorithm therefore involves tensions between several ethical ideals which have to be brought into an acceptable balance.

The transparency of an algorithm is further complicated by machine learning algorithms which are even more difficult to interpret and comprehend as they move along their learning process.[11] It is argued that the opacity of machine learning algorithms inhibits oversight. According to one scholar, algorithms are opaque in the sense that the recipient of an algorithm's output rarely has any concrete sense of how and why a particular classification has arrived at from inputs.[12] The opacity in machine learning algorithms appears to be a product of the high-dimensionality of data, complex code and changeable decision-making logic.[13] Therefore, it is further argued that meaningful oversight in algorithmic decision-making appears impossible when the machine has an informational advantage over the human operator.[14]

[8] Ananny (2015, p. 103).

[9] Glenn and Montieth (2014, p. 6).

[10] Leese (2014, p. 502).

[11] Burell (2016, p. 4), Leese (2014, p. 502), Hildebrandt (2011, pp. 378–380), Tutt (2017, p. 94).

[12] Burell (2016, p. 1).

[13] Burell (2016, p. 6).

[14] Matthias (2004, p. 182).

Even concerning algorithms operating on individually "hand-written" decision-making rules, it is argued that such algorithms can still be highly complex and practically inscrutable despite their lack of machine learning.[15] Especially when algorithms are developed by large teams of engineers over time, they cannot be divorced from the conditions under which they are developed, and this means that algorithms need to be understood as relational, contingent and contextual in nature and framed within the wider context of their sociotechnical assemblage.[16] Nevertheless, algorithmic processing contrasts with traditional human decision-making because the rationale of an algorithm may well be incomprehensible to humans which renders the legitimacy of its decisions difficult to challenge.[17]

Against this background, algorithmic decision-making hardly appears transparent, and opacity seems to prevent meaningful risk assessment. In the context of an elaborate automated tool, it would therefore currently appear rather unethical to have any action triggered by this tool other than further scrutiny of the gathered evidence by a human officer. Especially when the officer is aware of the ethical risks and dangers involved with the use of algorithms, the problem of scrutability seems suitably kept at bay.

7.6 Risk of Potential Bias

Within the literature reviewed for this chapter, the automation of human decision-making may not be justified by an alleged lack of bias in algorithms.[18] An algorithm's design and functionality reflects the values of its designer(s) and intended uses, if only to the extent that a particular design is preferred as the best or most efficient option.[19] Because the development of an algorithm involves many choices between several possible options, the values of the algorithm's author(s) are woven into the code which in effect institutionalises those values.[20] Without knowledge of the algorithm's development history, it is most difficult to detect latent bias in an algorithm.[21]

[15] Kitchin (2017, pp. 20 et seq).
[16] Kitchin (2017, p. 18).
[17] Mittelstadt et al. (2016, p. 7).
[18] Kitchin (2017, p. 18), Newell and Marabelli (2015, p. 6).
[19] Kitchin (2017, p. 18).
[20] Macnish (2012, p. 158).
[21] Hildebrandt (2011, p. 377).

In the context of an automated tool, it is also relevant that the outputs of algorithms require interpretation. Concerning behavioural data, the correlations presented by the algorithm might come to reflect the interpreter's unconscious motivations, socioeconomic determinations and geographic or demographic influences.[22] Therefore, a LEA officer evaluating the evidence presented by the automated tool has to be trained and aware that meaning is not self-evident in statistical models and that the explanation of any correlation requires additional justification. Different metrics make visible aspects of individuals and/or groups that are not otherwise perceptible.[23] Consequently, it may not be assumed that the LEA officer's interpretation of the evidence correctly reflects the perception of a targeted individual or group rather than the biases of the interpreter.

7.7 Unfair Discrimination

Whereas bias is a dimension of the decision-making process itself, an algorithm also creates the risk of leading to unfair discrimination based on an algorithm's profiling. The algorithm infers a pattern by means of data mining and thereby constructs a profile[24] which inevitably leads to discrimination if based on biased evidence decision-making process. An individual is comprehended based on connections with others identified by the algorithm, rather than based on actual behaviour.[25]

7.8 Unfairness of Algorithms

In the context of an automated tool, the risk of discrimination may emanate from a selector which is unreasonably based on prejudice about the likely characteristics of terrorist attackers. The choice of selectors might be too broad so that they single out a group of people on the basis of a trait which is not correlated with terrorism. Alternatively, the selectors might disproportionately identify communications of particular kinds of groups or individuals as suspicious who would then suffer from such indirect discrimination.

---

[22] Hildebrandt (2011, p. 376).

[23] Lupton (2014, p. 859).

[24] So the broad definition by Hildebrandt and Koops (2010, p. 431).

[25] Newell and Marabelli (2015, p. 5).

There appear to be four overlapping strategies for preventing such discrimination in general[26]:

1. Controlled distortion of training data
2. Integration of antidiscrimination criteria into the classifying algorithm
3. Post-processing of classification models
4. Modification of predictions and decisions to maintain a fair proportion of effects between protected and unprotected groups

These strategies are seen in the development of fairness-aware data mining which aims to be aware not only of discrimination but also fairness, neutrality and independence.[27] According to this approach of data mining, various metrics of fairness should already be integrated in the algorithm based on statistical parity, differential privacy and other parameters.[28]

7.9 Negative Effects on Societal Trust and Cohesion

The ethical acceptability of counterterrorist measures may be costly due to its effects on society at large. Especially surveillance measures more often than not affect individuals without any criminal record at a time when no crime has (yet) been committed. This raises the moral risk of social trust and cohesion being eroded by uses of technology.

7.10 Erosion of Trust in Policing Authorities

First, there is the citizen's trust in the policing authorities which could be weakened by what is perceived as excessive and ethically problematic use of technology (English Terrorism: How to Respond 2009). In a democracy, citizens are supposed to be allowed unobserved space in which to conduct their relationships and governments, and its agents are exposed to the scrutiny of those who are ruled which is a condition of the justified exercise of democratic power. Covert action by state institutions like LEAs

[26] Romei and Ruggieri (2014) as cited in Mittelstadt et al. (2016, p. 8).
[27] Mittelstadt et al. (2016, p. 8)
[28] Mittelstadt et al. (2016, p. 8).

could encourage citizens to doubt a central promise of democracy, namely, that the will of the people will be carried out by the people's institutions. Covert surveillance measures make it difficult to realise whether the will of the people is being done or not. However, in exceptional cases also covert action can be justified especially if the measure is taken to prevent great and imminent harm to citizens, and it is reasonable to believe that there is no public alternative available to LEAs at the time when they have to act. Further, even such covert actions require informing and getting permission from bodies under democratic control before a LEA engages in such covert operation.

7.11 Erosion of the Right to Be Trusted

Second, there is the citizen's right to be trusted as an expression of the more general presumption of innocence. This right to be trusted as innocent and norm-abiding citizen could be tainted if a surveillance measure was based on the premise that everybody is untrustworthy implying to some extent a presumption of guilt instead of presuming people innocent in the absence of evidence to the contrary. The right to be trusted is based on the ethical consideration that failure to presume people innocent of norm-breaking behaviour is incompatible with respect for them as moral agents (Duff 2013). However, the claim that failure to actively trust equates to active mistrust is a fallacy because it mistakenly assumes that trust and distrust are the only two trust-related attitudes it is possible to adopt. In fact, they appear to exist at opposite ends of a spectrum of attitudes (Ullmann-Margalit 2002). It would be equally unreasonable to require police to treat all individuals for whom there is no individually incriminating evidence of wrongdoing as if there existed evidence of their innocence with respect to the criminal law. However, asserting a right to be trusted implies that that is precisely what morality does require.

7.12 Discrimination of Minorities

Third, ethnic or political minorities may be stigmatised as terrorists. The risk for such minorities to be stigmatised as likely terrorists derives from the absence of a comprehensive definition of terrorism. The challenge of creating a comprehensive definition of terrorism condemning all terrorist activities has to do with the definitory precision required to permit the prosecution of criminal activities without condemning acts that should be deemed to be legitimate. Due to major divergences at the international level on the question of the legitimacy concerning the use of violence for political purposes, either by states or by self-determination and revolutionary groups, this has not yet been possible (Diaz-Paniagua 2008). Europe has experienced the effects of such divergences in the past, for example, when suffering from separatist groups like the IRA in Northern Ireland and the ETA in Spain. A more recent example may be the Kurdish minority in Syria or political Islamists. If, for example, the idea that Islamists can play a useful democratic role is broken, then they fall under a blanket repressing any Islamists which seems the worst possible response (Blanked repression is the wrong way to deal with political Islamists 2017). Islamists neither are all the same nor per se fundamentally undemocratic. Groups like Ennahda in Tunisia share power with secular groups because the fragile democratic transition requires broad consensus (Blanked repression is the wrong way to deal with political Islamists 2017). Ideologies must enjoy freedom to compete, as long as they abjure violence and respect democratic norms. Achieving a nonviolent political dialogue appears heavily influenced by whether the minority is stigmatised as terrorist association or not. Though such stigmatisation also seems influenced by the shared political view, peaceful and non-stigmatising political approaches are conceivable as demonstrated by the 2014 referendum in Scotland for independence from the UK.

References

M. Ananny, Toward an ethics of algorithms: Convening, observation, probability and timeliness. Sci. Technol. Hum. Values **41**(1), 93 (2015)

S. Barocas, Data mining and the discourse on discrimination, p. 2 under section 2.3 on "faulty inferences". Available at: https://dataethics.github.io/proceedings/DataMiningandtheDiscourseOnDiscrimination.pdf

Blanked repression is the wrong way to deal with political Islamists, The Economist, 26 August 2017. Available at: https://www.economist.com/news/leaders/21727067-their-record-power-often-worrying-they-can-be-pragmatic-and-cannot-be-ignored-blanket?frsc=dg%7Ce

J. Burell, How the machine thinks: Understanding opacity in machine learning algorithms. Big Data Secur. **3**(1), 1 (2016)

C.F. Diaz-Paniagua, *Negotiating Terrorism: The Negotiation Dynamics of Four UN Counter-Terrorism Treaties, 1997–2005* (City University of New York, New York, 2008), p. 47

P. Domingos, *The Master Algorithm: How the Quest for the Ultimate Learning Machine Will Remake the World* (Basic Books, New York, 2015)
A. Duff, Who must presume whom to be innocent of what? Neth. J. Leg. Philos. **42**(3), 170 (2013)
English, Terrorism: How to Respond, 2009, p. 141
T. Glenn, S. Montieth, New measures of mental state and behavior based on data collected from sensors, smartphones, and the internet. Curr. Psychiatry Rep. **16**(12), 1 (2014)
M. Hildebrandt, Who needs stories if you can get the data? Philos. Technol. **24**(4), 371 (2011)
M. Hildebrandt, B.J. Koops, The challenges of ambient law and legal protection in the profiling era. Mod. Law Rev. **73**(3), 428 (2010)
R.K. Hill, What an algorithm is. Philos. Technol. **29**(1), 35 (2016)
R. Kitchin, Thinking critically about and researching algorithms. Inf. Commun. Soc. **20**(1), 14 (2017)
M. Leese, The new profiling: Algorithms, black boxes, and the failure of anti-discriminatory safeguards in the European Union. Secur. Dialogue **45**(5), 494 (2014)
D. Lupton, The commodification of patient opinion: The digital patient experience economy in the age of big data. Sociol. Health Illn. **36**(6), 856 (2014)
K. Macnish, Unblinking eyes: The ethics of automating surveillance. Ethics Inf. Technol. **14**(2), 152 (2012)
A. Matthias, The resonsibility gap: Ascribing responsibility for the action of learning automata. Ethics Inf. Technol. **6**, 175 (2004)
B.D. Mittelstadt, P. Allo, M. Taddeo, S. Wachter, L. Floridi, The ethics of algorithms: Mapping the debate. Big Data Soc. **3**(2), 1 (2016)
S. Newell, M. Marabelli, Strategic opportunities (and challenges) of algorithmic decision-making: A call for action on the long-term societal effects of 'datafication'. J. Strateg. Inf. Syst. **24**(1), 3 (2015)
A. Romei, S. Ruggieri, A mulitdisciplinary survey on discrimination analysis. Knowl. Eng. Rev. **29**(5), 582–638 (2014)
E. Siegel, *Predictive Analytics: The Power to Predict Who Will Click, Buy, Lie, or Die* (John Wiley & Sons, 2016) Hoboken, New Jersey
R. Turner, N. Angius, The philosophy of computer science, in *The Stanford Encyclopedia of Philosophy*, ed. E.N. Zalta (2017). Available at: https://plato.stanford.edu/archives/spr2017/entries/computer-science/
A. Tutt, An FDA for algorithms. Adm. Law Rev. **69**, 83 (2017)
E. Ullmann-Margalit, Trust out of distrust. J. Philos. **99**(10), 532 (2002)

CHAPTER 8

Ethical and Societal Issues of Automated Dark Web Investigation: Part 3

Ulrich Gasper

8.1 Protection of Vital Interests of Citizens

The use of an automated tool by LEAs also has to be discussed as democratic expression of the will of the people. The automated tool requires a LEA to target a citizen or citizens it is supposed to protect. Such targeting or systematic monitoring of persons, places, and items in order to detect terrorist-specific conduct and to enable preventive or reactive measures involves large-scale collection of data and qualifies as a form of surveillance. Any form of surveillance not only questions the value of private life but also carries the risk of inhibiting the expression of the will of the people and the exercise of democratic rights such as the freedom to oppose a government. A liberal democratic government has no a prerogative to watch people who are peacefully minding their own business because it is in the interest of citizens not to be observed by LEAs when pursuing lawful personal projects. The interests governments are supposed to protect are those of the citizens they represent. It follows from this that the

U. Gasper (✉)
Cybercrime Research Institute, Cologne, Germany
e-mail: gasper@cybercrime.de

B. Akhgar et al. (eds.), *Dark Web Investigation*,
Security Informatics and Law Enforcement,
https://doi.org/10.1007/978-3-030-55343-2_8

permissibility by the norms of democracy depends on the element of consent by the citizens. Because an automated tool will be used covertly, collecting a citizen's direct informed consent is not an option. Rather, the permissibility of using an automated tool depends on the general consent of citizens to LEAs using lawful means of preventing the encroachments on the interests of citizens.

The use of an automated tool is a measure of surveillance as counterterrorism measure. The fight against terrorism enjoys a very strong prima facie claim to serving vital interests of citizens. Though "terrorism" could be defined opportunistically by governmental institutions, the fight against terrorism is well established in the European Union as principally a national competence. The European Union supports the efforts of Member States in several ways:

In April 2017, the Directive on Combatting Terrorism[1] entered into force strengthening the EU's legal framework in preventing terrorist attacks by criminalising acts such as providing[2] and receiving[3] training for terrorism and travel for terrorist purposes,[4] as well as organising or facilitating such travel.[5] The Directive on Combating Terrorism also reinforces the rights for the victims of terrorism.[6] Concerning the use of automated tools, Article 20(1) Directive on Combating Terrorism encourages Member States to make "effective investigative tools" (such as those which are used in organised crime or other serious crime cases) also available to LEAs investigating or prosecuting the terrorist offences defined in Articles 3–12 Directive on Combating Terrorism.

In May 2017, a new Regulation on Europol entered into force and took effect in all EU Member States enabling Europol to step up efforts

[1] Directive (EU) 2017/541 of the European Parliament and of the Council of 15 March 2017 on Combating Terrorism, Official Journal of theEU, L 88/6 v. 31 March 2017.

[2] Article 7 Directive (EU) 2017/541 of 15 March 2017 on Combating Terrorism, Official Journal of the EU, L 88/6 v. 31 March 2017.

[3] Article 8 Directive (EU) 2017/541 of 15 March 2017 on Combating Terrorism, Official Journal of the EU, L 88/6 v. 31 March 2017.

[4] Article 9 Directive (EU) 2017/541 of 15 March 2017 on Combating Terrorism, Official Journal of the EU, L 88/6 v. 31 March 2017.

[5] Article 10 Directive (EU) 2017/541 of 15 March 2017 on Combating Terrorism, Official Journal of the EU, L 88/6 v. 31 March 2017.

[6] Articles 24, 25 and 26 Directive (EU) 2017/541 of 15 March 2017 on Combating Terrorism, Official Journal of the EU, L 88/6 v. 31 March 2017.

to fight terrorism and establishing Europol as the European Union Agency for Law Enforcement Cooperation with a view to supporting cooperation among law enforcement authorities in the Union.[7]

The EU counterterrorism strategy[8] aims to combat terrorism regionally and beyond while respecting human rights. At the heart of all political efforts is making Europe safer and allowing its citizens to live in an area of freedom, security, and justice. The safety and survival of individuals and the interest in nonviolent collective self-determination are vital interests of the EU citizens which are threatened by terrorism.

8.2 Monitoring an Automated Tool's Effectiveness

An automated tool meant to be used for preventive counterterrorism activities may include large-scale collection of data by LEAs which is performed covertly. In such scenario, the automated tool appears as highly intrusive technology especially because it could be applied to people against whom there is either no evidence of wrongdoing at all or merely less than compelling evidence. Although the potential harm of an act of terrorism is very high, at the time a LEA uses the automated tool neither the probability of this harm is established to be high nor a very likely source for it might be established. As a consequence, the effectiveness of using the automated tool appears in doubt, while the probability of intruding deeply and unjustifiably into the lives of individuals who are not involved in terrorism seems rather high.

Against this background, LEAs using such an automated tool will have to monitor the tool's effectiveness so that they will have an evidence base from which to draw for future decisions about its use in operations.

---

[7] Regulation (EU) 2016/794 of the European Parliament and of the Council of 11 May 2016 on the European Union Agency for Law Enforcement Cooperation (Europol), Official Journal of the EU, 24 May 2016, L 135/53.

[8] Council of the EU, The European Union Counter-Terrorism Strategy, No. 14469/5/05 REV 4, 30 November 2005.

8.3 Finding Suitable Criteria as Selectors Incorporated in an Automated Tool

If a sophisticated automated tool is used by a LEA once sufficient evidence for a reasonable suspicion of terrorism acts against an individual has been established, the already existing suspicion against the individual should justify the attempt to broaden the evidence base for the suspicion with the help of the automated tool.

A different and very challenging question is whether and how the automated tool could establish without any prior suspicion any reliable evidence for a reasonable suspicion against an individual to be somehow involved with acts of terrorism. Such automated tool may be envisaged to search, crawl, and monitor online spaces and forums for content relevant for terrorist activities. In this respect, the automated tool could be viewed as searching for patterns supposedly characteristic of terrorism also in spaces and forums with perfectly innocent online activities.

One possibility of focusing an automated tool on genuinely pertinent information could be the automated filtering of information on the basis of selectors which would then automatically exclude the vast majority of material from intrusive further inspection by a LEA. However, there are various difficulties intertwined with the use of selectors.

8.4 Particular Keywords as Selector

First, there is the possibility of choosing particular keywords as selectors. An ideal selector keyword in this regard would be a word which is known to be used exclusively in the context of terrorist activities, perhaps something like a secret codeword. Next best would be words providing reason for suspicion based strongly on evidence (such as highly specialist explosive materials or weapons). Other possible keywords appear to be rather terms used by large numbers of people for almost any reason (e.g. "terrorism"). A keyword may not be discriminatory because it has to be indicative only of suspicious terrorist activity. A single keyword appears difficult to define in this regard, but also a set of several keywords may not contain a discriminatory keyword because the use of certain words (e.g. "muslim") is not only ambivalent but also more likely to be used by certain religious groups and in that respect discriminatory. It follows from all this that a keyword used as selector for an automated tool should be reasonable, evidence-based, and nondiscriminatory. In addition, it has to be born in mind that suitable selectors have to be flexible enough to respond to suspects who are forensically aware and aim to avoid the use of incriminating language.

8.5 Names of Specific Groups as Selectors

Second, the names of specific groups might serve as good selectors: It is possible to find names of specific groups providing a stronger evidence base for terrorist purposes. Not many people may have come across, e.g. "El Waliki" or "Al Nusra". Such specific group names, however, could either gain a prominence in the press for whatever reason or the specific group is involved in legitimate political life.

8.6 Names of Particular Individuals as Selectors

Third, it may appear that names of an individual known to be a terrorist could be a good selector. However, searches for names of such individuals are quite likely to produce an unacceptably large number of false-positives. Behind each false-positive is an innocent individual who coincidentally shares its name with a suspicious individual. Further, matching the name of a suspicious individual in a database appears prone to error in itself and especially when a name in foreign script is transcribed into the Roman alphabet (Branting 2005). Error can hardly be neutral and most likely directs suspicion and scrutiny at members of cultural groups because names are to a large extent culturally inherited. At the very least, the number of false-positives has to be factored into any assessment of proportionality of the intrusion by the automated tool.

8.7 Useful Considerations for Appropriateness of Selectors

Against this background, the appropriateness of a selector for the automated tool should be determined by at least the following considerations:

- How likely are terrorists and their associates to be using those terms?
- What is the ratio between such suspicious potentially terrorist users of the specific group name and the innocent people using the same term?

- How easy is it to distinguish between suspicious potentially terrorist users and innocent users of those terms once the selection has been made?

8.8 Avoidance of Chilling Effects

The use of covert surveillance has to be accounted for in democratic societies. The mere knowledge about the use of covert surveillance tools by LEAs may lead citizens to a certain wariness as to how such tools may be employed. Fearing the inconvenience of being detained on a certain suspicion and then released without any conviction, citizens might be disinclined from engaging in otherwise perfectly legitimate online activities.[9] This frame of mind may further be intimidated by the power gained by the surveillant over the surveilled and may potentially cause an undesirable self-censorship leading to a loss of spontaneity when online. Such "chilling effects" are at odds with democratic values and practice. Drawing on the presumption of innocence, no reason should be given for such chilling effects.

One potential reason involving such effects is a stigmatisation as criminally suspicious. On an individual level, being stigmatised as having failed to maintain the moral standards of the community can be humiliating. Such humiliation is not intended when following up on a suspicion, but it can often be a side effect of such suspicion even if the interference by a LEA is proportionate and well founded. If such individuals perceive the suspicion to be an unjust implication of wrongdoing, then this may create knock-on social costs by not only eroding their trust in LEAs but also reducing their willingness to cooperate with LEAs.[10] Stigmatisation may also make those affected feel alienated which could damage their self-confidence. When particular groups of people who share salient traits (e.g. religion or race) are stigmatised as suspicious, this may either intensify already existing prejudices against them[11] or even create new prejudices

---

[9] See Stoicheff (2016); the Washington Post reported about this study under the Headline "Mass surveillance silences minority opinions", Washington Post, March 28, 2016. See also the PEN American Center's study "Chilling Effects: NSA Surveillance Drives U.S. Writers to Self-Censor" published November 12, 2013 presenting Research conducted by The FDR Group (thefdrgroup.com/).

[10] See section "non-reporting of discrimination" in: Fundamental Rights Agency of the EU (FRA), (Report) Respect for and protection of persons belonging to minorities 2008–2010, pp. 38 and 39.

[11] See chapter Race Law and Suspicion, in: Kennedy, Race, Crime and the Law, 1997; Lever, Why racial profiling is unjustified, (2004) Philosophy and Public Affairs, 32(2).

(O'Connor and Rumann 2003). All these potential social costs are likely to increase along with the importance of the crime one is suspected of having committed. These harms of stigmatisation and their chilling effects impose a moral duty on LEAs to refrain from stigmatising people as criminally suspicious without any good enough reason. It appears generally accepted that evidence linking a specific individual to a particular future or past crime is a sufficient ground for treating the individual as a suspect and inflicting on them the costs of such treatment.

For the use of an automated tool by LEAs, it seems a valid evaluation that the stronger the evidence is, the more justified appears the use of highly stigmatising measures of suspicion. Further, as long as a LEA follows procedures that ensure that the surveillance will be stopped as soon as it becomes clear that insufficient evidence exists for continued suspicion, the measure could be defended as proportionate and ethically legitimate. Placing suspicion on innocent people behaving in such a way as to fit a profile for affiliation to terrorist activities is undeserved but not ethically unfair if inflicted only to the extent proportionate and necessary to fight against terrorism. The right not to be stigmatised as suspicious has to be balanced against the need for LEAs to have sufficient powers at their disposal to be able to prevent and investigate terrorist activities. These powers must be sufficiently broad to allow LEAs to cast a net wide enough to catch terrorists and to pursue tentative leads.

8.9 Assessing Surveillance Technologies

Finally, it has to be pointed out that a nuanced approach to assessing surveillance technologies has been elaborated by the collaborative project SURVEILLE from the EU's Seventh Framework Programme for research and technological development.[12] The project SURVEILLE has developed a comprehensive methodology for everyone including police authorities to determine whether it is legal, moral, efficient, and effective to use a particular surveillance technology (Assessing surveillance technologies 2016). The methodology not only takes into account the impact of a surveillance technology on fundamental rights like the right to privacy and freedom of expression but measures also effectiveness including cost and can be applied to a wide range of situations including when deciding about the development or deployment of new surveillance technologies.[13] It is,

[12] SURVEILLE – Surveillance. Ethical Issues, Legal Limitations and Efficiency, see: https://surveille.eui.eu

[13] "Assessing surveillance technologies", SURVEILLE briefing note, February 2016, p. 2.

therefore, suggested and recommended to have an automated tool for LEAs evaluated according to this comprehensive methodology.

References

Assessing surveillance technologies, SURVEILLE briefing note (2016 February). Available at: https://surveille.eui.eu/wp-content/uploads/sites/19/2015/07/SURVEILLE-Policy-Brief.pdf

L.K. Branting, Name matching in law enforcement and counter-terrorism (2005). Available at: http://www.karlbranting.net/papers/icail2005.pdf

M.P. O'Conno, C.M. Rumann, Into the fire: How to avoid getting burned by the same mistake made fighting terrorism in Northern Ireland. Cordozo Law Rev. **24**, 1657 (2003)

E. Stoicheff, Under surveillance: Examining Facebook's spiral of silence effects in the wake of NSA internet monitoring. Journal. Mass Commun. Q. **93**(2), 296–311 (2016)

CHAPTER 9

Ethical and Societal Issues of Automated Dark Web Investigation: Part 4

Marco Gercke

9.1 Introduction

There are areas that should specifically be paid attention to during the development of an automated tool, such as LEA's need to be aware that despite harmonisation approaches the legislation legal framework criminalising terrorist activities differs from country to country and for any LEA the use of any automated tool has to be based on a legitimate legal basis which includes the use of an automated tool independently of a specific case.

If different LEAs are involved in antiterrorism operations, the mandate of each LEA to use a sophisticated automated tool needs to be accessed individually, and certain functions might need to be restricted for some, and using a tool that is able to automate searches and collect large volumes of data may not be covered by provisions authorising the search and collection of data in a manual fashion. When implementing undercover operation capabilities, it is important to design it in accordance with national legal frameworks and practices.

M. Gercke (✉)
Cybercrime Research Institute, Cologne, Germany
e-mail: gercke@cybercrime.de

B. Akhgar et al. (eds.), *Dark Web Investigation*,
Security Informatics and Law Enforcement,
https://doi.org/10.1007/978-3-030-55343-2_9

The collection of large volumes of data may include personal data and consequently raise questions with regard to data protection. Cross-border collection of information raises questions related to national sovereignty, and exchanging information cross border might fall under formal agreements related to international cooperation in criminal matters.

During the collection of terrorist content, illegal content might be collected. The possession of such material could lead to criminal investigations, and collecting and storing media that was published online could violate copyright laws. While social media analysis could idenitfy perpetrators, collecting information from social media platforms could at the same time violate licensing or end user agreements. If the data collected through an automated tool should be used as evidence in court, it will be important to ensure that the rules and regulations with regard to the collection of evidence are observed when designing the tool.

9.2 Legal Issues for Consideration

9.2.1 Harmonised Criminal Law Framework as Foundation for Cooperation

For centuries, criminal law and criminal procedural law were a domain of national legislation. Criminal law is a main area influenced by history and culture.[1] Despite similarities and harmonisation approaches by United Nations (see Chap. 10), Council of Europe (see Chap. 10) and European Union (see Chap. 10), criminal law and crime-related policies still vary from country to country.[2]

However, unlike prior inventions like electricity (where different countries use different voltages and different plugs), the Internet is based on

[1] See: Herlin-Karnell (2007a), page 69 et seq; *Hecker*, Sind die nationalen Grenzen des Strafrechts ueberwindbar? Die Harmonisierung des materiellen Strafrechts in der Europaeischen Union, JA 2007, page 561 et seq; Herlin-Karnell (2007b), page 15 et seq; *Rosenau*, Zur Europaeisierung des Strafrecht, ZIS 2008, page 9 et seq; Ambos (2005), page 173 et seq; *Nuotio*, Criminal Law and Cultural Sensitivity, Refaerd Argang 31, 2008, Nr. 1/120, page 18; *Johnstone/Jones*, History of Criminal Justice, 2011, page 6; Siegel von Wadsworth, Criminology: Theories, Patterns, and Typologies, 2012, page 7.

[2] See in this regard: The Criminal Law Competence of the European Union, House of Lords, London, HL Paper 227, 2006; Yakut (2009), page 1; Gercke (2010)

single technical standard.[3] Any country that ignores fundamental protocols would de facto risk to be disconnected from the global network. The need to respect global standards is not limited to technology but relevant for legislation as well. While it is theoretically possible for a country to develop legislation addressing terrorist use of the Internet and related investigation based on dogmatic concepts that significantly differ from global trends and best practices, such approach would limit the country's ability to participate in the global fight against terrorism. A harmonised legal framework is of great importance as quite a few countries base their mutual legal assistance regime on the principle of "dual criminality". Dual criminality exists if the offence is a crime under both the requested and requesting party's laws.[4] Investigations on a global level are in this case limited to those crimes that are criminalised in all cooperating countries. If countries – based on the paradigm that criminal law is national domain – develop standards that differ from international best practices, this can hinder the ability of international cooperation and ultimately lead to safe havens.[5] Harmonisation of legislation is, consequently, identified as a key priority by different regional and international organisations.[6]

[3] Regarding technical standardization, see: OECD (2007). Regarding the importance of single technical as well as single legal standards, see: Gercke (2008) page 7 *et seq.*

[4] The difficulties the dual criminality principle can cause within international investigations are a current issue in a number of international conventions and treaties. Examples include Art. 2 of the EU Framework Decision of 13 June 2002 on the European arrest warrant and the surrender procedures between Member States (2002/584/JHA). Regarding the dual criminality principle in international investigations, see: United Nations Manual on the Prevention and Control of Computer-Related Crime, 269, available at http://www.uncjin.org/Documents/EighthCongress.html; Schjolberg and Hubbard (2005); *Plachta*, International Cooperation in the Draft United Nations Convention against Transnational Crimes, UNAFEI Resource Material Series No. 57, 114th International Training Course, page 87 *et seq.*, available at: http://www.unafei.or.jp/english/pdf/PDF_rms/no57/57-08.pdf

[5] The issue has been addressed by a number of international organizations. UN General Assembly Resolution 55/63 stipulates: "States should ensure that their laws and practice eliminate safe havens for those who criminally misuse information technologies". The full text of the resolution is available at: http://www.unodc.org/pdf/crime/a_res_55/res5563e.pdf. The G8 Ten-Point Action Plan highlights: "There must be no safe havens for those who abuse information technologies".

[6] Declaration Synergies and Responses: Strategic Alliances in Crime Prevention and Criminal Justice, available at: http://www.unodc.org/pdf/crime/congress11/BangkokDeclaration.pdf

9.2.2 *Consideration for Using an Automated Tool Cross Border*

Harmonised legislation is of great importance when it comes to cross-border cooperation of LEAs. Improving such cooperation would be one relevant aim developing a sophisticated automated tool for use by LEAs across countries. Both the European Union and the Council of Europe have undertaken various approaches to harmonise the legislation in the member states. This eases cooperation among LEAs of different member states. It is, however, important to underline that on the EU-level this was carried out through Framework Decisions and Directives and not through Regulations. Based on Art. 288 Treaty of the Functioning of the European Union, only the results to be achieved through Directives are binding but shall leave to the national authorities the choice of form and methods. Consequently, there are still differences when it comes to the criminalisation of terrorist activities. Therefore, it appears recommendable that priority is given to terrorist activities where legislation is aligned.

> Existence of legislation authorising the search for terrorist content independently of an individual case.

At least in civil law countries, LEAs will not be able to investigate crimes without specific laws in place authorising such investigation.[7] In order to carry out the investigations, they need to be able to base their investigations on procedural instruments that enable them to take the measures that are necessary to identify an offender and collect the evidence required for the criminal proceedings.[8] These measures can be the same ones that are undertaken in other investigations not related to Internet-related content. However, investigating activities of criminals or members of terrorist organisations that act online goes along with some unique challenges. As a consequence, investigations may be carried out in a different way

[7] This was also highlighted by the drafters of the Council of Europe Convention on Cybercrime, which contains a set of essential investigation instruments. The drafters of the report point out: "Not only must substantive criminal law keep abreast of these new abuses, but so must criminal procedural law and investigative techniques", see: Explanatory Report to the Council of Europe Convention on Cybercrime, No. 132.

[8] Regarding user-based approaches in the fight against cybercrime, see: Goerling (2006). See also the comment made by *Jean-Pierre Chevenement*, French Minister of Interior, at the G8 Conference in Paris in 2000: "More broadly, we have to educate users. They must all understand what they can and can't do on the Internet and be warned of the potential dangers. As use of the Internet grows, we'll naturally have to step up our efforts in this respect".

compared to traditional investigations.[9] If an offender is, for example, based in one country,[10] uses services that enable anonymous communication and, in addition, published terrorist content online by using different public Internet terminals, the identification of the suspect can hardly be based on traditional instruments like search and seizure alone.

With regard to criminal investigations related to terrorist use of the Internet, the Council of Europe Convention on Cybercrime contains a set of provisions that reflect widely accepted minimum standards regarding procedural instruments required for online investigations.[11] The Convention even addresses issues of high relevance – such as the question if LEAs are allowed to access information available on servers located in another country.

While this is very helpful in criminal investigations, it is important to underline that providing automated tools to LEAs does not solely focus on enhancing the ability of LEAs in investigating crime. The need for automation has a strong focus on prevention. As a consequence, the intended content-based functions of the tool – such as searching, crawling, monitoring and gathering of data – are not part of a specific investigation but in general independent of investigations in a specific case.

9.2.3 *Consideration for Using an Automated Tool Cross Border*

When it comes to the implementation and utilisation of automated tools across countries, it will be important for LEAs to clarify one main question: Is there a provision in national legislation that permits LEAs to carry out automated searches for potential terrorist content independently of a specific case.

---

[9] Due to the protocols used in Internet communication and worldwide accessibility, there is very little need for a physical presence at the place where a service is physically offered. Due to this independence of place of action and the crime site, many criminal offences related to the Internet are transnational crimes.

[10] The pure fact that the offender is acting from a different country can result in additional challenges for law-enforcement agencies' investigations even if similar substantive criminal law provisions and procedural law instruments are in place in both countries. In these cases, the investigation nevertheless requires international cooperation between the authorities in both countries, which in general is more time consuming compared to investigations concentrating on a single country.

[11] See Articles 15–21 of the Council of Europe Convention on Cybercrime.

9.3 Responsible Authority (Antiterrorism, Law Enforcement, National Security Intelligence Services) and their Legal Basis

Closely related to the issue discussed in Sect. 9.2.2 (automated search independently of a specific case) is the question which authority is responsible for using the different functions provided by an automated tool. In most countries, not all LEAs are automatically mandated to search for terrorist content online. Some countries strictly separate between preventive police work and investigation of crimes. In addition, some countries have specialised units that are responsible for dealing with terrorism.

9.3.1 Consideration for Using an Automated Tool Cross Border

When it comes to the implementation and utilisation of automated tools in countries with various authorities and split up mandates, it could be important to be able to restrict specific functions of the tool. Such restriction could be achieved through a customisation.

9.4 Impact of Automation

A review of the legislation in different EU Member States reveals that various investigative instruments exist, that cover the collection of information. However, it is important to underline that the impact of single action (carried out by an investigator) and an automated mass collection of data could have a significantly different impact.[12] Depending on the capacities of the tool used to collect data from public sources, the tool will not only be able to identify specific content but even produce profiles of people that published content. The automation of such data collection for criminal investigations was contested in different member states. In 2008, the German Constitutional Court, for example, decided that the automatic collection of registered license plate information on motorways can be permitted, but storing such data for longer periods of time can go along

[12] With regard to the impact of mass surveillance see for example: *Lyon*, Surveillance, power, and everyday life, published in Avgerou/Mansell/Quah/SilverStone, The Oxford Handbook of Information and Communication Technologies, 2009; Surveillance and censorship: The impact of technologies on human rights, European Parliament, Directorate-General for External Policies, 2015.

with a violation of fundamental rights related to privacy.[13] Storing the collected data is a potential main function of an automated tool.

9.4.1 *Consideration for Using an Automated Tool*

When identifying provisions authorising the use of an automated tool, it is important to not only focus on provisions authorising a single and manual act but the automated collection of data.

9.5 Undercover Operations and Agent Provocateur

Some functions of an automated tool (such as automatic communication in forums where terrorist content is posted and discussed) generally require covert operations. Such interaction could fail if the tool used identities that reveal that they are associated to a LEA. Within an end user assessment, end users point out their need for being able to create social media accounts that cannot be traced back to the acting LEA. Undercover Internet investigations are not a new technique. They are, for example, used with regard to Internet sex offenders.[14] The use of "agent provocateur" is an instrument also used in investigations related to online child abuse.[15] The use in investigations related to terrorism is intensively discussed, and various cases are documented.[16]

9.5.1 *Consideration for Using an Automated Tool Cross Border*

In some EU Member States, the application of such advanced investigation instruments as part of an automated tool can be challenging. Consequently, it could be important to be able to restrict specific functions of the tool. Such restriction could be achieved through a customisation.

---

[13] See: German Constitutional Court, 1 BvR 2075/05. With regard to decisions from different EU member states see: Kindt (2013), page 927 et seq.

[14] See for example: Mitchell et al. (2011); Vendius (2015), page 6 et seq.

[15] See Martellozzo (2015), page 32 et seq.

[16] See for example: *Bjelopera*, The Federal Bureau of Investigation and Terrorism Investigations, 2013, CRS R 41780.

9.6 Data Protection

With the continuing development of new technologies, data protection is a topic of increasing relevance. Apple's, Google's and Facebook's advertising policies, the NSA-scandal, cross-border flow of data and the heated discussions about "the right to be forgotten": it seems that the call for the protection of data and privacy has never been more present.

However, just defining what privacy is is challenging.[17] Some state privacy is an important fundamental human right as it underpins human dignity and other values such as freedom of association and freedom of speech.[18] However, the introduction of new technologies challenges privacy as they facilitate the collection, storage, processing and combination of personal at a larger scale and faster pace than ever.

The link to using an automated tool is that the large-scale collection of data is one of its key components. This data collection will most likely include personal data. It will be challenging to differentiate here at the time of collection as ultimately everything is code – whether it is a computer virus, a political speech or a video published by a terrorist organisation. Content filtering (as required by anti-child pornography and pro-intellectual property lobbies) requires deep packet inspection, which is extremely intrusive from the point of view of privacy and data protection.[19] Any automated tool might face similar challenges.

9.6.1 Consideration for Using an Automated Tool Cross Border

It will be important to ensure that an automated tool is designed in a way that it guarantees the protection of personal data and privacy in line with binding laws both on the European Level and the Member States. In this regard, especially the EU General Data Protection Regulation is of great relevance and should serve as guiding principle. However, at the same time even stricter national legislation should be reflected.

[17] *Svantesson*, A Legal Method for Solving Issues of Internet Regulation; Applied to the Regulation of Cross-Border Privacy Issues. EUI Working Papers LAW No. 2010/18.

[18] Friedewald et al. (2010), page 61 et seq.

[19] Porcedda, Maria Grazia (2012), Data Protection and the Prevention of Cybercrime: The EU as an area of security? Via: http://cadmus.eui.eu/handle/1814/23296

9.7 Cross-Border Collection and Exchange of Information

Investigating crimes with a cross-border dimension requires a close cooperation between LEAs in all the countries affected.[20] Cross-border investigations undertaken without the consent of the competent authorities of the affected countries may violate the fundamental principle of national sovereignty. This principle prohibits countries to carry out investigations within the territory of another country without the permission of the local authorities.[21]

Bilateral agreements and multilateral agreements such as the United Nations Convention against Transnational Organised Crime (UNTOC)[22] and its three protocols,[23] the Inter-American Convention on Mutual Assistance in Criminal Matters[24] and the European Convention on Mutual Assistance in Criminal Matters[25] provide solutions for key issues. With Europol, the EU Member States have an institutional framework for expedited exchange of information and coordination of investigations.

The Convention on Cybercrime addresses this issue in Art. 32:

---

[20] Regarding the need for international cooperation in the fight against cybercrime, see: *Putnam/Elliott*, International Responses to Cyber Crime, in *Sofaer/Goodman*, The Transnational Dimension of Cyber Crime and Terrorism, 2001, page 35 *et seq.*, available at: http://media.hoover.org/documents/0817999825_35.pdf; Sofaer/Goodman, Cyber Crime and Security – The Transnational Dimension in *Sofaer/Goodman*, The Transnational Dimension of Cyber Crime and Terrorism, 2001, page 1 *et seq.*, available at: http://media.hoover.org/documents/0817999825_1.pdf

[21] National sovereignty is a fundamental principle in international law. See *Roth*, State Sovereignty, International Legality, and Moral Disagreement, 2005, page 1, available at: http://www.law.uga.edu/intl/roth.pdf

[22] Convention Against Transnational Organized Crime (2000), GA RES/55/25, Entry into Force: 29.09.2003. Regarding the Convention, see: Smith (2009).

[23] The Protocol to Prevent, Suppress and Punish Trafficking in Persons, Especially Women and, the Protocol against the Smuggling of Migrants by Land, Sea and Air and the Protocol Against the Illicit Manufacturing of and Trafficking in Firearms, Their Parts and Components and Ammunition.

[24] Inter-American Convention on Mutual Assistance in Criminal Matters, 1992, Treaty Series, OAS, No. 75. The text of the Convention and a list of signatures and ratifications is available at: http://www.oas.org/juridico/english/sigs/a-55.html

[25] European (Council of Europe) Convention on Mutual Assistance in Criminal Matters, 1959, ETS 30.

Article 32 – Trans-border access to stored computer data with consent or where publicly available

A Party may, without the authorisation of another Party:

a access publicly available (open source) stored computer data, regardless of where the data is located geographically; or

[...]

Especially when it comes to a more informal exchange of information, for example, in a kind of social media platform closed to LEAs, binding legal standards are rare. One example for spontaneous information exchange outside the traditional cooperation is Art. 26 Convention on Cybercrime. However, there are concerns related to an increasing informal information sharing substituting formal international cooperation.[26]

9.7.1 Consideration for Using an Automated Tool Cross Border

Information exchange is one fundamental component of a sophisticated automated tool. If the idea is to provide a platform for planning, this would meet a high demand of the LEA community to exchange cross border. However, during the design process of such an automated tool, it will be important to precisely focus on those areas where such cooperation does not contradict a more formal cooperation. In addition, national security aspects need to be taken into consideration when it comes to cross-border collection of data.

9.8 Illegal Content

Without doubt, the Internet is used to make available, trade and exchange illegal content. This ranges from child pornography, xenophobic material or insults related to terrorist content.[27] With regard to illegal content, value systems and legal systems differ extensively between countries.[28] The dissemination of xenophobic material may be illegal in many European

[26] Gercke (2014), page 294.

[27] For reports on cases involving illegal content, see *Sieber*, Council of Europe Organised Crime Report 2004, page 137 *et seq.*

[28] Gercke (2014), page 23.

countries,[29] but protected by the principle of freedom of speech[30] in others like the United States.[31] The use of derogatory remarks with respect to the Holy Prophet is criminal in many Arabic countries, but not in some European countries.[32] A criminalisation of illegal content should not interfere with the right to freedom of expression as, for example, defined by principle 1 (b) of the Johannesburg Principles on National Security and Freedom of Expression.[33] However, principle 1 (c) clarifies that the right to freedom of expression may be subject to restrictions.

But despite the possibility to restrict freedom of expression, it is important to underline that based on the concerns mentioned above, a restriction has to be strictly limited. Such limitations are especially discussed with regard to the criminalisation of defamation.[34] The 2008 Joint Declaration of the UN Special Rapporteur on Freedom of Opinion and Expression

[29] One example of the wide criminalisation of illegal content is Sec. 86a German Penal Code. The provision criminalises the use of symbols of unconstitutional parties: Section 86a: Use of Symbols of Unconstitutional Organizations.

[30] Regarding the principle of freedom of speech, see: *Tedford/Herbeck/Haiman*, Freedom of Speech in the United States, 2005; *Barendt*, Freedom of Speech, 2007; *Baker*, Human Liberty and Freedom of Speech; *Emord*, Freedom, Technology and the First Amendment, 1991. Regarding the importance of the principle with regard to electronic surveillance, see: Woo and So (2002), page 530 *et seq.*; *Vhesterman*, Freedom of Speech in Australian Law; A Delicate Plant, 2000; Volokh (2001), page 57 *et seq.*; *Cohen*, Freedom of Speech and Press: Exceptions to the First Amendment, CRS Report for Congress 95–815, 2007.

[31] Concerns over freedom of expression (e.g. the First Amendment to the United States Constitution) explain why certain acts of racism were not made illegal by the Council of Europe Convention on Cybercrime, but their criminalisation was included in the First Additional Protocol. See Explanatory Report to the Council of Europe Cybercrime Convention First Additional Protocol, No. 4.

[32] The 2006 Joint Declaration of the UN Special Rapporteur on Freedom of Opinion and Expression, the OSCE Representative on Freedom of the Media and the OAS Special Rapporteur on Freedom of Expression points out that "in many countries, overbroad rules in this area are abused by the powerful to limit non-traditional, dissenting, critical, or minority voices, or discussion about challenging social issues". In 2008 the Joint Declaration highlights that international organizations, including the United Nations General Assembly and Human Rights Council, should desist from the further adoption of statements supporting the idea of defamation of religions.

[33] 1996 Johannesburg Principles on National Security, Freedom of Expression and Access to Information.

[34] The 2002 Joint Declaration of the UN Special Rapporteur on Freedom of Opinion and Expression, the OSCE Representative on Freedom of the Media and the OAS Special Rapporteur on Freedom of Expression points out that "defamation is not a justifiable restriction on freedom of expression; all criminal defamation laws should be abolished and replaced, where necessary, with appropriate civil defamation laws".

and others points out that vague notions such as providing communications and the glorification or promotion of terrorism or extremism should not be criminalised.[35]

With regard to some content (such as child pornography), the degree of consent to criminalise acts related to such material is wider. Child pornography is, for example, broadly condemned, and offences related to child pornography are widely recognised as criminal acts.[36] International organisations are engaged in the fight against online child pornography,[37] with several international legal initiatives, including the 1989 United Nations Convention on the Rights of the Child,[38] the 2011 European Union Directive on combating the sexual abuse and sexual exploitation of children and child pornography[39] and the 2007 Council of Europe Convention on the Protection of Children against Sexual Exploitation and Sexual Abuse, among others.[40] Approaches to criminalising child pornography are designed in general to protect different legal interests. Criminalisation of the production of child pornography seeks to protect children from falling victim to sexual abuse.[41] However, it is important to highlight that most approaches to criminalise child pornography go beyond criminalising the production or dissemination of such material. The 2011 European Union Directive on combating the sexual abuse and sexual exploitation of children and child pornography that defines the standard for the EU Member States does not only contain a criminalisation of possession but also criminalises obtaining access by

---

[35] International Mechanisms for Promoting Freedom of Expression, Joint Declaration on Defamation of Religions, and Anti-Terrorism and Anti-Extremism Legislation, by the UN Special Rapporteur on Freedom of Opinion and Expression, the OSCE Representative on Freedom of the Media and the OAS Special Rapporteur on Freedom of Expression, and the ACHPR (African Commission on Human and Peoples Rights) Special Rapporteur on Freedom of Expression and Access to Information, 2008.

[36] ITU Global Cybersecurity Agenda / High-Level Experts Group, Global Strategic Report, 2008, page 34.

[37] See, for example, the "G8 Communique", Genoa Summit, 2001.

[38] United Nations Convention on the Right of the Child, A/RES/44/25, Regarding the importance of cybercrime legislation see: ITU Global Cybersecurity Agenda / High-Level Experts Group, Global Strategic Report, 2008.

[39] Directive 2011/92/EU of the European Parliament and of the Council of 13 December 2011 on combating the sexual abuse and sexual exploitation of children and child pornography, and replacing Council Framework Decision 2004/68/JHA.

[40] Council of Europe Convention on the Protection of Children against Sexual Exploitation and Sexual Abuse, CETS No: 201.

[41] *Levesque* (1999), page 68.

means of information and communication technology, to child pornography.

There are two aspects that make the discussion about criminalising illegal content highly relevant for the use of automated tools by LEAs:

- *First of all*, the tool aims to enable LEAs to search and crawl both the surface and dark web for terrorist-generated content. This can take place for the purpose of collecting information about group dynamics or planning of real-world attacks. However, the publication of some of the content may by itself be illegal – even if it does not lead to a real-world attack. The prosecution of offences related to illegal content could in some cases be easier than to prove the involvement in a broader plot. If the latter fails, it could be interesting for LEAs to have the ability to prosecute on the basis of illegal information. To support this endeavour, a filter that identifies illegal material and therefore allows to separate this material from other content could be helpful.
- The *second* implication for the use of an automated tool is a potential criminal liability for material collected. It is possible that the tool collects illegal material. Already in 2008, there were reports that terrorists used child pornography websites to exchange information.[42] Some criminal law systems do explicitly decriminalise acts undertaken by law enforcement officials. One example is Sec. 184b, paragraph 5 of the German Penal Code. While Sec. 184b paragraph 1–4 criminalise acts related to child pornography (from distribution to possession), paragraph 5 states that the criminalisation of possession shall not apply to acts that exclusively serve the fulfilment of lawful officials or professional duties. While this decriminalisation certainly applies to officers working in the fight against child pornography, it may not automatically apply to law enforcement officers that work in the fight against terrorism. Furthermore, there might be an additional challenge if such material is made available to other LEAs in another country. Very often this will happen unintentionally and is, therefore, not criminalised. However, the tool might need to address this issue.

[42] Tibbetts, Terrorists use child porn to exchange information, The Telegraph, 10.10.2008.

9.8.1 *Consideration for Using an Automated Tool*

The collection and exchange of information are fundamental components of most automated tools. Within the design of the tool, special consideration should be given to the fact that the collection and exchange of information may include illegal material, and despite harmonisation approaches, the criminal law systems of the member states show differences when it comes to the criminalisation of illegal content and preventing law enforcement officials from being prosecuted for interacting or exchanging such material.

9.9 Copyright Issues

Whenever large quantities of information are collected through crawlers, it raises questions related to the material. One issue is illegal content that could be among the data that the crawler collects (see Sect. 9.8 above). Another issue frequently discussed is copyright violation. The crawler might copy and save content in a database that is protected by copyright laws. This issue was frequently discussed with regard to search engines.[43] Search engines often use technology similar to the one most likely utilised in sophisticated automated tools. From a legal point of view, it would, however, be too easy to point out that it can hardly be illegal for law enforcement to do what search engines do on a daily basis. There are two main concerns related to such argumentation:

- *First of all*, some countries have specifically addressed the liability of search engines. Search engines play an import role in the successful development of the Internet, and it was quite rightly pointed out that "without much exaggeration one could say that to exist is to be indexed by a search engine".[44] The European Union E-Commerce Directive[45] does not contain standards defining the liability of search

[43] See in this regard for example: *Rotenberg/Compano*, Search Engines for Audio-Visual Content: Copyright Law and its Policy Relevance, published in Preissl et al., Telecommunication Markets, 2009.

[44] *Introna/Nissenbaum*, Sharping the Web: Why the politics of search engines matters, page 5.

[45] Directive 2000/31/EC of the European Parliament and of the Council of 8 June 2000 on certain legal aspects of information society services, in particular, electronic commerce, in the Internal Market (Directive on electronic commerce).

engine operators. Therefore, some EU countries have decided to address the liability of search engine providers in a dedicated provision.[46] Unlike in the case of hyperlinks, not all countries have based their regulation on the same principles.[47] Spain[48] and Portugal have based their regulations regarding the liability of search engine operators on Article 14 of the Directive, while Austria[49] has based the limitation of liability on Article 12. This framework cannot simply be transferred to crawlers utilised by LEA.

- Second, there are some countries that have specifically addressed the issue of copyright violations within the context of regular work of LEAs. One example is Art. 45 of the German Copyright Act. It

---

[46] Austria, Spain and Portugal. See Report of the application of the Directive on electronic commerce – COM (2003) 702, page 7.

[47] See Report of the application of the Directive on electronic commerce – COM (2003) 702, page 15.

[48] Ley de Servicios de la Sociedad de la Información y de Comercio Electrónico (LSSICE) – Artículo 17. Responsabilidad de los prestadores de servicios que faciliten enlaces a contenidos o instrumentos de búsqueda (Spain)

1. Los prestadores de servicios de la sociedad de la información que faciliten enlaces a otros contenidos o incluyan en los suyos directorios o instrumentos de búsqueda de contenidos no serán responsables por la información a la que dirijan a los destinatarios de sus servicios, siempre que: a) No. tengan conocimiento efectivo de que la actividad o la información a la que remiten o recomiendan es ilícita o de que lesiona bienes o derechos de un tercero susceptibles de indemnización, o b) si lo tienen, actúen con diligencia para suprimir o inutilizar el enlace correspondiente.

Se entenderá que el prestador de servicios tiene el conocimiento efectivo a que se refiere la letra a) cuando un órgano competente haya declarado la ilicitud de los datos, ordenado su retirada o que se imposibilite el acceso a los mismos, o se hubiera declarado la existencia de la lesión, y el prestador conociera la correspondiente resolución, sin perjuicio de los procedimientos de detección y retirada de contenidos que los prestadores apliquen en virtud de acuerdos voluntarios y de otros medios de conocimiento efectivo que pudieran establecerse.

2. La exención de responsabilidad establecida en el apartado primero no operará en el supuesto de que el destinatario del servicio actúe bajo la dirección, autoridad o control del prestador que facilite la localización de esos contenidos.

[49] Ausschluss der Verantwortlichkeit bei Suchmaschinen § 14. (1) Ein Diensteanbieter, der Nutzern eine Suchmaschine oder andere elektronische Hilfsmittel zur Suche nach fremden Informationen bereitstellt, ist für die abgefragten Informationen nicht verantwortlich, sofern er

1. die Übermittlung der abgefragten Informationen nicht veranlasst,
2. den Empfänger der abgefragten Informationen nicht auswählt und
3. die abgefragten Informationen weder auswählt noch verändert.

(2) Abs. 1 ist nicht anzuwenden, wenn die Person, von der die abgefragten Informationen stammen, dem Diensteanbieter untersteht oder von ihm beaufsichtigt wird.

authorises courts and authorities to make copies of copyright protected material for the purpose of administration or justice and public security.

9.9.1 Consideration for Using an Automated Tool Cross Border

The collection of large volumes of data by an automated tool – especially when the tool contains capacities for media analysis – goes along with potential copyright conflicts. Within the design of the tool, special consideration should be given to an operation with the legal basis defined by law in the member states.

9.10 Licensing and End User Agreement Issues

Closely related to the issue of copyright are issues related to terms and conditions for using websites that potentially contain illegal material. A lot of platforms are freely accessible. However, copying all content or even just parts does raise questions with regard to not only potential copyright violations but also potential violations of the terms and conditions of the operator of the platform. Very often the "End User Agreement" contains provisions that restrict the ability to collect and use data from public profiles. In 2010, an entrepreneur who collected 2010 million datasets from Facebook through a crawler faced legal threats by the company.[50] It is important to underline that various social media companies are supporting the work of LEAs. However, the fact that they often created special portals for law enforcement and have implemented routines makes it unlikely that they will necessarily tolerate the mass collection of data by tools such as TENSOR.

9.10.1 Consideration for Using an Automated Tool

When it comes to the collection of data from social media platforms, working on the basis of existing interfaces and established procedures should be taken into consideration. When using crawlers, the tool should avoid violations of license agreements be mindful about the fact that most popular services are run by countries outside the EU.

---

[50] *Giles*, Data sifted from Facebook swiped after legal threats, New Scientist, 31.03.2010.

9.11 Electronic Evidence

Ideally, the information collected by an automated tool could serve as evidence when an offender is prosecuted. In cases where no traditional evidence is available, the ability to successfully identify and prosecute an offender may be based on the rightful collection and evaluation of digital evidence.[51] However, this goes along with unique challenges when it comes to designing the process of collecting and processing data[52]:

One of the most fundamental requirements for the admissibility of both traditional categories of evidence[53] and digital evidence alike is, for example, the legitimacy of evidence.[54] This fundamental principle requires that digital evidence has been collected, analysed, preserved and finally presented in court in accordance with the appropriate procedures and without violating the fundamental rights of the suspect.[55] Both the requirements relating to the collection, analysis, preservation and finally presentation of the evidence in court and the consequences of a violation of the suspect's rights differ from country to country. Principles and rules that can possibly be violated range from fundamental rights of a suspect such as privacy[56] to failure to respect procedural requirements.[57] Due to the often inadequate legislation, general principles of evidence are frequently applied to digital evidence.[58]

Another example is the best evidence rule that is of great relevance for common-law jurisdictions.[59] There are some references, mostly in old cases, to a "best evidence rule", which under common law provides that only the best available evidence of a fact at issue is said to be admissible.

---

[51] Regarding the need for formalization of computer forensics, see: Leigland and Krings (2004).

[52] Regarding the challenges related to handling digital evidence see: Casey (2004), page 9.

[53] Regarding the legitimacy principle, see: *Grans/Palmer*, Australian Principles of Evidence, 2005, page 10.

[54] *Insa*, Situation Report on the Admissibility of Electronic Evidence in Europe, in: Syllabus to the European Certificate on Cybercrime and E-Evidence, 2008, page 219.

[55] *Malaga*, Requirements for the Admissibility in Court of Digital Evidence, in: Syllabus to the European Certificate on Cybercrime and E-Evidence, 2008, page 207.

[56] Winick (1994), page 80.

[57] Gercke (2014), page 249.

[58] Malaga, Requirements for the Admissibility in Court of Digital Evidence, in: Syllabus to the European Certificate on Cybercrime and E-Evidence, 2008, page 208.

[59] *Kenneally*, UCLA Journal of Law and Technology, 2005, Vol. 9, Issue 2; *Keane*, Modern Law of Evidence, 2005, page 27.

The general rule now appears to be that whether a given item of evidence is the best available evidence or not only affects its weight, not its admissibility.[60]

Closely related to the best evidence rule, the "primary evidence rule" formerly provided that in the case of documentary evidence, only the original document or an "enrolled" copy of that document was admissible to prove its contents and authenticity. With regard to digital evidence, this raises a number of questions, insofar as it is necessary to determine what the original is.[61]

These are only three examples of the principles that may need to be respected if collected information should serve as evidence in court.

9.11.1 *Consideration for Using an Automated Tool Cross Border*

If it is intended to use data collected through an automated tool in court, special attention should be given to ensure that the data is collected and processed in line with the requirement of the respective country where the tool is utilised.

References

K. Ambos, Is the development of a common substantive criminal law for Europe possible? Some preliminary reflections. Maastricht J. Eur. Compar. Law **12**, 173 (2005)

E. Casey, *Digital Evidence and Computer Crime* (Academic Press, Cambridge, 2004)

M. Friedewald, D. Wright, S. Gutwirth, E. Mordini, Privacy, data protection and emerging sciences and technologies: Towards a common framework – innovation. Eur. J. Soc. Sci. Res. **23**(1), 61 (2010)

M. Gercke, National regional and international approaches in the fight against cybercrime. Comp. Law Rev. Int. **9**, 7 (2008)

M. Gercke, Impact of the Lisbon treaty on fighting cybercrime in the EU. Comp. Law Rev. Int. **11**, 75 (2010)

M. Gercke, *Understanding Cybercrime* (ITU, Geneva, 2014), p. 23

S. Goerling, The Myth of User Education, 2006., Available at: www.parasite-economy.com/texts/StefanGorlingVB2006.pdf

---

[60] Halsbury's Laws of England, Vol. 11(3): Criminal Law, Evidence and Procedure, 2006, pages 331–332.

[61] *Clough*, The Admissibility of Digital Evidence, 2002.

E. Herlin-Karnell, Commission v. Council: Some reflections on criminal law in the first pillar. Eur. Public Law **13**, 69 (2007a)

E. Herlin-Karnell, Recent developments in the area of European criminal law. Maastricht J. Eur. Compar. Law **14**, 15 (2007b)

E.J. Kindt, *Privacy and Data Protection Issues of Biometric Applications* (Springer, Dordrecht, 2013), p. 927

R. Leigland, A.W. Krings, A formalization of digital forensics. Int. J. Digit. Evid. **3**(2), 1–32 (2004)

R.J. Levesque, Sexual Abuse of Children: A Human Rights Perspective (Indiana University Press, Cambridge, 1999), p. 68

E. Martellozzo, Policing online child sexual abuse – The British experience. Eur. J. Pol. Stud. **3**(1), 32 (2015)

K.J. Mitchell, J. Wolak, D. Finkelhor, L. Jones, Investigators using the internet to apprehend sex offenders: Findings from the Second National Juvenile Online Victimization Study. Police Pract. Res. 13(3), 1–15 (2011)

OECD, Internet Address Space, Economic Consideration in the Management of IPv4 and in the Development of IPv6 (2007), DSTI/ICCP(2007)20/FINAL. Available at: http://www.itu.int/dms_pub/itu-t/oth/06/15/T061500000A0015PDFE.pdf

J.S. Schjolberg, A.M. Hubbard, Harmonizing national legal approaches on cybercrime, 2005, p. 5. Available at: http://www.itu.int/osg/spu/cybersecurity/presentations/session12_schjolberg.pdf

J.M. Smith, An international hit job: Prosecuting organized crime acts as crimes against humanity. Georgetown Law J. **97**, 1118 (2009). Available at: http://www.georgetownlawjournal.org/issues/pdf/97-4/Smith.PDF

T.T. Vendius, Proactive undercover policing and sexual crimes against children on the internet. Eur. Rev. Org. Crime **2**(2), 6 (2015)

E. Volokh, Freedom of speech, religious harassment law, and religious accommodation law. Loyola Univ. Chicago Law J. **33**, 57 (2001)

R. Winick, Search and seizures of computers and computer data. Harv. J. Law Technol. **8**(1), 80 (1994)

C. Woo, M. So, The case for Magic Lantern: September 11 Highlights the need for increasing surveillance. Harv. J. Law Technol. **15**(2), 530 (2002)

B. Yakut, Post-Lisbon criminal law competences of the European Union. Marmara J. Eur. Stud. **17**, 1–52 (2009)

CHAPTER 10

Ethical and Societal Issues of Automated Dark Web Investigation: Part 5

Ulrich Gasper

10.1 Introduction

An automated tool most likely is intended for searching the Internet for potential terrorist-related content independently of an individual case. For that purpose, officers of a LEA enter a suspicious name, place or activity into an automated tool which then searches the Internet and the dark web and presents as search results in what ways the searched (id)entity appears connected to terrorist activities and to which other suspicious groups or identities it can be traced.

This chapter provides an overview of the relevant legal framework at global (Sect. 10.2) and at European (Sects. 10.3 and 10.4) level for the use of such an automated tool by LEAs.

U. Gasper (✉)
Cybercrime Research Institute, Cologne, Germany
e-mail: gasper@cybercrime.de

B. Akhgar et al. (eds.), *Dark Web Investigation*,
Security Informatics and Law Enforcement,
https://doi.org/10.1007/978-3-030-55343-2_10

10.2 Global Legal Framework

Terrorism is an act against human rights which gives states not only the right but also the obligation to do something against it. For states, the respect for human rights and the rule of law is an integral part of the fight against terrorism. The applicable global international legal framework related to counterterrorism is contained in a range of sources, including resolutions of the General Assembly and the Security Council, treaties, jurisprudence and customary international law.

10.2.1 Universal Declaration of Human Rights and the Rule of Law

The international community has committed to adopting measures that ensure respect for human rights for all and the rule of law as the fundamental basis of the fight against terrorism, through the adoption of the United Nations Global Counter-Terrorism Strategy by the General Assembly in its 2006 Resolution 60/288.[1] According to the "plan of action" in the Annex of UN Resolution 60/288, the Member States of the United Nations have resolved to take measures aimed at addressing the conditions conducive to the spread of terrorism and ensure that any measures taken to counterterrorism comply with their obligations under international law, in particular, human rights law, refugee law and international humanitarian law.[2]

The United Nations Global Counter-Terrorism Strategy reaffirms the inextricable links between human rights and security and places respect for the rule of law and human rights at the core of national and international counterterrorism efforts.[3] Through this Strategy, the Member States have committed to ensuring respect for human rights and the rule of law as the fundamental basis of the fight against terrorism recognising, in particular,

[1] United Nations, General Assembly, Resolution 60/288, "The United Nations Global Counter-Terrorism Strategy", A/RES/60/288, adopted on 8 September 2006.

[2] No. 3 of Annex 'Plan of action' to United Nations, General Assembly, Resolution 60/288, A/RES/60/288, 8 September 2006, and No. 1 of United Nations, General Assembly, Resolution 60/158, "Protection of human rights and fundamental freedoms while countering terrorism", A/RES/60/158, adopted on 16 December 2005.

[3] United Nations, General Assembly, Resolution 60/288, A/RES/60/288, 8 September 2006, Annex 'Plan of Action', part I. and in particular part IV. referring in No. 1 to Resolution 60/158 by United Nations General Assembly, A/RES/60/158, 16 December 2005.

that "effective counter-terrorism measures and the protection of human rights are not conflicting goals, but complementary and mutually reinforcing".[4] To be effective, this should include the development of national counterterrorism strategies that seek not only to prevent acts of terrorism and address the conditions conducive to their spread but also to prosecute or lawfully extradite those responsible for such criminal acts and to give due attention to the rights of all victims of human rights violations. In particular, concerning measures to prevent and combat terrorism, the Member States resolved to explore ways and means to (a) coordinate efforts to counterterrorism in all its forms and manifestations on the Internet and (b) use the Internet as a tool for countering the spread of terrorism while recognising that States may require assistance in this regard.[5]

Due care must be taken to respect international human rights standards in all phases of counterterrorism initiatives, including preventive gathering of intelligence and evidence. This requires the development of national counterterrorism legislation and practices that promote and protect fundamental human rights and the rule of law.

Based on the United Nations Global Counter-Terrorism Strategy, effective counterterrorism measures and the protection of human rights are complementary and mutually reinforcing objectives which must be pursued together. Counterterrorism initiatives relating to Internet use may have an impact on the enjoyment of a range of human rights, including the right to freedom of expression (Art. 19 UDHR), the right to freedom of association (Art. 20 UDHR), the right to privacy (Art. 12 UDHR) and the right to a fair trial (Art. 10, 11 UDHR).[6] None of these human rights belongs to the list of non-derogable rights and freedoms under Art. 4(2) of the International Covenant of Civil and Political Rights (ICCPR). Therefore, these human rights may be subject to narrowly construed limitations and derogations.

[4] Introduction of part IV. of Annex 'Plan of action' to United Nations, General Assembly, Resolution 60/288, A/RES/60/288, 8 September 2006.

[5] No. 12. of part II. of Annex "Plan of action' to United Nations, General Assembly, Resolution 60/288, A/RES/60/288, 8 September 2006.

[6] United Nations, General Assembly, Universal Declaration of Human Rights (UDHR) Resolution 217 A, A/RES/3/217 A, 10 December 1948.

10.2.2 Conditions for Limiting Human Rights

It seems important to highlight that the vast majority of counterterrorism measures are adopted on the basis of ordinary national legislation. In a limited set of exceptional national circumstances, some restrictions on the enjoyment of certain human rights may be permissible.

As provided for by international human rights conventions, Member States may legitimately limit the exercise of certain so-called qualified rights including the right to freedom of expression,[7] the right to freedom of association and assembly,[8] the right to freedom of movement[9] and the right to respect for one's private and family life.[10] In order to fully respect their human rights obligations while imposing such limitations, Member States must demonstrate the necessity of a limitation and may only take such measures as are proportionate to the pursuance of legitimate aims in order to ensure continuous and effective protection of human rights.[11] Consequently, in addition to respecting the principles of equality and non-discrimination, the limitations must be prescribed by law, in pursuance of one or more specific legitimate purposes and proportionate. In no case may the limitations be applied or invoked in a manner that would impair the essence of a human right.[12]

Taking the right to freedom of expression[13] as an example, it is not an absolute right and may be restricted when that freedom is (mis)used to incite discrimination, hostility or violence. Any restriction is subject to satisfaction of strictly construed tests of legality, necessity, proportionality and nondiscrimination. However, a key difficulty in cases of glorification of or incitement to terrorism is identifying where the line of acceptability lies, because this varies greatly from country to country depending on differing cultural and legal histories. Concerning antiterrorism measures another key difficulty is that restricting freedom of expression in response

[7] Art. 19 UDHR = Art. 19 ICCPR.

[8] Art. 20 UDHR = Art. 21 and 22 ICCPR.

[9] Art. 13 UDHR = Art. 12 ICCPR.

[10] Art. 12 UDHR = Art. 17 ICCPR.

[11] No. 6 of United Nations, International Covenant of Civil and Political Rights, General Comment No. 31 [80] on "The Nature of the General Legal Obligation Imposed on States Parties to the Covenant" adopted on 29 March 2004.

[12] No. 6 of United Nations, International Covenant of Civil and Political Rights, General Comment No. 31 [80] on "The Nature of the General Legal Obligation Imposed on States Parties to the Covenant" adopted on 29 March 2004.

[13] Art. 19 UDHR = Art. 19 ICCPR.

to terrorism might facilitate certain terrorist objectives which, in particular, include the dismantling of human rights protection.[14]

10.2.3 Gathering of Suspicious Information by LEAs

Countering terrorist use of the Internet may involve the surveillance and collection of information relating to suspected individuals. Due regard should be given to protecting persons against arbitrary or unlawful interference with the right to privacy enshrined in Art. 12 UDHR. The right to privacy includes the right to privacy of information about an individual's identity as well as his or her private life.

Domestic laws must be sufficiently detailed regarding, inter alia, the specific circumstances in which such interference may be permitted. Appropriate safeguards must also be in place to prevent abuse of secret surveillance tools. Further, any personal data collected must be adequately protected against unlawful or arbitrary access, disclosure or use.

10.2.4 Due Process Rights

Guaranteeing due process rights is critical for ensuring that counterterrorism measures are effective and respect the rule of law. Human rights protections for anyone charged with terrorism-related criminal offences include the right to be presumed innocent (Art. 11 UDHR), the right to a hearing with due guarantees and within a reasonable time by a competent, independent and impartial tribunal (Art. 10 UDHR) and the right to have a conviction and sentence reviewed by a higher tribunal that meets the same standards.[15]

10.2.5 Framework of UN Resolutions

The United Nations plays a pivotal role in developing an integrated response to terrorism across borders and among national criminal justice systems. Security Council resolutions may impose legally binding

---

[14] See, International Mechanisms for Promoting Freedom of Expression", Joint Declaration by UN Special Rapporteur on Freedom of Opinion and Expression, the OSCE Representative on Freedom of the Media and the OAS Special Rapporteur on Freedom of Expression, 21 December 2005.

[15] United Nations, Human Rights, Terrorism and Counter-terrorism, Fact Sheet No. 32, 2008, at part III.F. on p. 38.

obligations on Member States or provide "soft law" sources of political commitments or emerging norms of international law. Council resolutions adopted under Chapter VII of the Charter of the United Nations are binding on all Member States. The General Assembly has also adopted a number of resolutions relating to terrorism which provide useful sources of soft law and have high political importance, even though they are not legally binding.

The gradual development of a normative and legal framework based on the United Nations Global Counter-Terrorism Strategy is documented by the Secretary-General in his most recent Report on implementing activities of the United Nations system after the Strategy's first decade.[16] Annex I of this 2016 Report lists all 19 international legal instruments which include the International Convention for the Suppression of the Terrorist Bombings 1997, the International Convention for the Suppression of the Financing of Terrorism 1999 and the International Convention for the Suppression of Nuclear Terrorism 2005.[17] The 2016 Report also mentions the General Assembly's 1994 Declaration on Measures to Eliminate International Terrorism[18] as milestone development and refers to 48 resolutions of the General Assembly addressing various aspects of terrorism such as the protection of human rights and fundamental freedoms while countering terrorism as well as mandates of specialised United Nations bodies (e.g. the United Nations Counter-Terrorism Implementation Task Force, UN CTITF and the United Nations Counter-Terrorism Centre, UNCCT).[19] Finally, the 2016 Report not only outlines five resolutions of the Security Council strengthening the legal framework for preventing and combating terrorism but also describes the following eight recent key resolutions of the Security Council[20]:

---

[16] United Nations, Report of the Secretary-General on "Activities of the United Nations system in implementing the United Nations Global Counter-Terrorism Strategy", A/70/826, 12 April 2016.

[17] Annex I of United Nations, Report of the Secretary-General on "Activities of the United Nations system in implementing the United Nations Global Counter-Terrorism Strategy", A/70/826, 12 April 2016.

[18] United Nations, General Assembly, "Declaration of Measures to Eliminate International Terrorism", A/RES/49/60, 9 December 1994.

[19] Annex I of United Nations, Report of the Secretary-General on "Activities of the United Nations system in implementing the United Nations Global Counter-Terrorism Strategy", A/70/826, 12 April 2016.

[20] Annex I of United Nations, Report of the Secretary-General on "Activities of the United Nations system in implementing the United Nations Global Counter-Terrorism Strategy", A/70/826, 12 April 2016.

- On kidnapping and hostage-taking committed by terrorist groups, SCR 2133 (2014)[21]
- On foreign terrorist fighters, terrorist financing, sanctions and reporting, SCR 2170 (2014) invoking Chapter VII[22]
- On flow of foreign fighters and financing or other support to terrorist groups and on countering violent extremism, SCR 2178 (2014) invoking Chapter VII[23]
- On preventing terrorist from benefiting from transnational organised crime, SCR 2195 (2014)[24]
- On preventing terrorist groups from benefiting from trade in oil, antiquities and hostages and from receiving donations, SCR 2199 (2015) invoking Chapter VII[25]
- On countering violent extremism and terrorism, SCR 2242 (2015)[26]
- On dismantling funding and support channels for terrorist groups, SCR 2253 (2015) invoking Chapter VII[27]

The Secretary-General's Report led to the General Assembly's Resolution 70/291 reviewing the first decade for the United Nations Global Counter-Terrorism Strategy.[28] In this most recent Review, the General Assembly:

- Stresses that counterterrorism efforts neglecting the rule of law, at the national and international levels, not only betray the values they

---

[21] United Nations, Security Council, Resolution 2133 (2014), S/RES/2133 (2014), 27 January 2014.

[22] United Nations, Security Council, Resolution 2170 (2014), S/RES/2170 (2014), 15 August 2014.

[23] United Nations, Security Council, Resolution 2178 (2014), S/RES/2178 (2014), 24 September 2014.

[24] United Nations, Security Council, Resolution 2195 (2014), S/RES/2195 (2014), 19 December 2014.

[25] United Nations, Security Council, Resolution 2199 (2015), S/RES/2199 (2015), 12 February 2015.

[26] United Nations, Security Council, Resolution 2242 (2015), S/RES/2242 (2015), 13 October 2015.

[27] United Nations, Security Council, Resolution 2253 (2015), S/RES/2253 (2015), 17 December 2015.

[28] United Nations, General Assembly, "The United Nations Global Counter-Terrorism Strategy Review", A/RES/70/291, 1 July 2016.

seek to uphold but also may further fuel violent extremism that can be conducive to terrorism[29]
- Urges to respect and protect the right to privacy in the context of digital communication[30]
- Urges Member States to provide full coordination and afford one another the greatest measure of assistance in criminal investigations or criminal proceedings relating to the financing or support of terrorist acts[31]
- Notes the urgent need for the international community to globally counter terrorists exploiting information and communications technologies, including through the Internet and social media, for crafting distorted narratives that justify violence and which are utilised to recruit supporters and foreign terrorist fighters, mobilise resources and garner support from sympathisers.[32]

In 2016, the Security Council added two resolutions relating to the fight against terrorism without invoking Chapter VII of the Charter of the United Nations:

- On countering terrorist threats to civil aviation, SCR 2309 (2016)[33]
- On criminalising the financing of terrorism in domestic law and on establishing national contact points for 24/7 network, SCR 2322 (2016)[34]

There is currently no comprehensive United Nations treaty on terrorism that is applicable to an exhaustive list of the manifestations of terrorism. Similarly, the international community has yet to agree on an internationally binding definition of the term "terrorism", owing largely

---

[29] No. 16 of United Nations, General Assembly, "The United Nations Global Counter-Terrorism Strategy Review", A/RES/70/291, 1 July 2016.

[30] No. 19 and No. 20 of United Nations, General Assembly, "The United Nations Global Counter-Terrorism Strategy Review", A/RES/70/291, 1 July 2016.

[31] No. 32 of United Nations, General Assembly, "The United Nations Global Counter-Terrorism Strategy Review", A/RES/70/291, 1 July 2016.

[32] No. 54 of United Nations, General Assembly, "The United Nations Global Counter-Terrorism Strategy Review", A/RES/70/291, 1 July 2016.

[33] United Nations, Security Council, Resolution 2309 (2016), S/RES/2309 (2016), 22 September 2016.

[34] United Nations, Security Council, Resolution 2322 (2016), S/RES/2322 (2016), 12 December 2016.

to the difficulty of devising a universally acceptable legal categorisation for acts of violence committed by States, by individuals or by armed groups such as liberation or self-determination movements.

The universal instruments do not define terrorist offences as crimes under international law. Rather, they create an obligation for States parties to the agreements to criminalise the specified unlawful conduct under their domestic law, exercise jurisdiction over offenders under prescribed conditions and provide for international cooperation mechanisms that enable States parties to either prosecute or extradite the alleged offenders. Until the successful conclusion of ongoing negotiations on a universal definition or comprehensive convention relating to terrorism, bilateral and multilateral agreements should provide the basis for the development of common standards to counter the use of the Internet for terrorist purposes, in the interest of promoting international cooperation.

No universal convention has been adopted specifically relating to the prevention and suppression of terrorist use of the Internet. Because anti-terrorism investigations online require transborder access to, copying of, search and seizure of electronic data, the G8 States have agreed on principles on transborder access to stored computer data.[35] According to these principles, the standard international procedure for getting hold of electronic data in another State is Mutual Legal Assistance (MLA). However, in two scenarios LEAs of a State do not have to obtain authorisation from the targeted State: First, when accessing "publicly available (open source) data" and, second, when acting in accordance with "the lawful and voluntary consent of a person who has the lawful authority to disclose that data".[36]

10.3 Regional Legal Framework: Council of Europe

At regional level, the Member States of the Council of Europe base their guarantee of human rights not only on the UDHR but also on the European Convention for the Protection of Human Rights and

[35] Annex I to Ministerial Conference of the G-8 Countries on Combating Transnational Organised Crime, Moscow, 19–20 October 1999.

[36] No. 6 of Annex I to Ministerial Conference of the G-8 Countries on Combating Transnational Organised Crime, Moscow, 19–20 October 1999.

Fundamental Freedoms[37] better known as the European Convention on Human Rights (ECHR). Because the ECHR was declared considering the Universal Declaration of Human Rights (UDHR), the ECHR comprises much the same guarantees of fundamental rights and freedoms as the UDHR.

10.3.1 *ECHR Guarantees and Their Legitimate Restrictions*

Relevant for the cross-border use of an automated tool are especially the guarantees of the right to a fair trial[38] based on the presumption of innocence,[39] the right to respect of private and family life (privacy)[40] as well as the freedom of expression[41] and the freedom of assembly and association.[42] It is crucial to note that the right to private and family life,[43] the freedom of expression[44] and the freedom of assembly and association[45] allow for legitimate restrictions in accordance with the law which are "necessary in a democratic society" in the interests of "national security", "public safety" or "for the prevention of disorder or crime" among other legitimate reasons. According to Article 18 ECHR, such legitimate restrictions may not be applied for any purpose other than those for which they have been prescribed.

To ensure the observance of the protection of human rights and fundamental freedoms, the ECHR established the European Court of Human Rights (ECtHR) in Articles 19–51 ECHR. The judgements of the ECtHR are binding for and enforceable in all Member States according to Article 46 ECHR.

Concerning the use of an automated tool for searches of the Internet and the dark web enabling LEAs to profile suspected individual in detail, the ECtHR has developed a catalogue of minimum standards for

---

[37] Convention for the Protection of Human Rights and Fundamental Freedoms (as amended by Protocols Nos. 11 and 14 and supplemented by Protocols Nos. 1, 4, 6, 7, 12 and 13), 4 December 1950.

[38] Art. 6 ECHR.

[39] Art. 6(2) ECHR.

[40] Art. 8 ECHR.

[41] Art. 10 ECHR.

[42] Art. 11 ECHR.

[43] Art. 8(2) ECHR.

[44] Art. 10(2) ECHR.

[45] Art. 11(2) ECHR.

surveillance which are crucial for the evaluation of national surveillance measures.[46]

This catalogue of minimum standards includes the following:

- The definition of the addressee of a national surveillance measure
- A list of crimes justifying a national surveillance measure
- A time limit for the national surveillance measure
- A procedure for the use of the collected data
- Measures ensuring that the collected data remains undamaged and available for subsequent control
- Provisions for the deletion of the collected data[47]

These minimum standards for surveillance were originally established for surveillance of telephone communications of individuals.[48] In 2008, the ECtHR transferred these minimum standards to all other surveillance measures on the basis that the potential negative effect on fundamental rights was identical.[49] The comparability of the negative effect on fundamental rights and freedoms also affects the type of data which is collected. The ECtHR acknowledged that the collection of content data is as problematic as the collection of traffic data concerning the intensity of encroaching fundamental rights and freedoms. In its more recent case law, the ECtHR specified these minimum standards on several occasions more precisely:

- Supportive material for measure: The ECtHR demanded a requirement in national legislation for the authorities to demonstrate the relation between the persons concerned by the surveillance measure

[46] ECtHR, decision of 4 December 2015 in case of *Zakharov v. Russia*, Application No. 47143/06, at para. 231 and ECtHR, decision of 12 January 2016 in case of *Szabo and Vissy v. Hungary*, Application No. 37138/14, at para. 66.

[47] ECtHR, decision of 30 July 1998 in case of *Valenzuela Contreras v. Spain*, Application No. 58/1997/824/1048, at para. 46; ECtHR, decision of 4 December 2015 in case of *Zakharov v. Russia*, Application No. 47143/06, at para. 231 and ECtHR, decision of 12 January 2016 in case of *Szabo and Vissy v. Hungary*, Application No. 37138/14, at para. 66.

[48] ECtHR, decision of 6 September 1978 in case of *Klass and others v. Germany*, Application No. 5029/71.

[49] ECtHR, decision of 1 July 2008 in case of *Liberty and Others v. UK*, Application No. 58243/00, at para. 63.

and the prevention of any terrorist threat.[50] There has to be a legal safeguard requiring LEAs to produce supportive materials or, in particular, a sufficient factual basis for the application of secret intelligence gathering measures which then enable the evaluation of necessity of the proposed measure - and this on the basis of an individual suspicion regarding the target person.[51]

- Effective control: Based on the rule of law, the ECtHR required that an interference by LEAs with an individual's rights should be subject to an effective control which should normally be assured by the judiciary, at least in the last resort, because judicial control offers the best guarantees of independence, impartiality and a proper procedure.[52]
- Acute privacy protection: In a more general way, the ECtHR drew from the technological advances the conclusion that the potential interferences of (mass) electronic surveillance with email, mobile phone and Internet services demanded the ECHR's protection of private life even more acutely.[53]
- Profiling: Because these data often compile further information and present LEAs with an opportunity for profiling, the ECtHR emphasised that the possibility for LEAs to acquire a detailed profile of the most intimate aspects of citizens' lives may result in particularly invasive interferences with private life and requires that this threat to privacy must be subjected to very close scrutiny both on the domestic level and under the ECHR.[54]
- Prevention of uncontrolled surveillance: The ECtHR demanded to prevent substituting a terrorist threat for a perceived threat of unfettered executive power intruding into citizens' private spheres by

---

[50] ECtHR, decision of 12 January 2016 in case of *Szabo and Vissy v. Hungary*, Application No. 37138/14, at para. 67.

[51] ECtHR, decision of 12 January 2016 in case of *Szabo and Vissy v. Hungary*, Application No. 37138/14, at para. 71; ECtHR, decision of 4 December 2015 in case of *Zakharov v. Russia*, Application No. 47143/06, at para. 260 .

[52] ECtHR, decision of 12 January 2016 in case of *Szabo and Vissy v. Hungary*, Application No. 37138/14, at para. 75 and 77.

[53] ECtHR, decision of 12 January 2016 in case of *Szabo and Vissy v. Hungary*, Application No. 37138/14, at para. 53.

[54] ECtHR, decision of 12 January 2016 in case of *Szabo and Vissy v. Hungary*, Application No. 37138/14, at para. 70.

virtue of uncontrolled yet far-reaching surveillance techniques and prerogatives.[55]

- Strict necessity of secret surveillance: Given the potential of cutting-edge surveillance technologies to invade citizens' privacy, the ECtHR expounded the requirement "necessary in a democratic society" in this context as requiring "strict necessity" in two aspects:

1. A measure of secret surveillance can be found as being in compliance with the ECHR only if it is strictly necessary, as a general consideration, *for the safeguarding the democratic institutions*
2. If it is strictly necessary, as a particular consideration, *for the obtaining of vital intelligence* in an individual operation[56]

The ECtHR established this safeguard to limit a LEA's discretion in interpreting the broad terms of "persons concerned identified" by following an established practice to verify whether sufficient reasons for intercepting a specific individual's communications exist in each case.[57]

10.3.2 European Convention on Cybercrime

In 2001, the Council of Europe (CoE) elaborated the Convention on Cybercrime which is still the only multilateral, legally binding instrument addressing criminal activity conducted via the Internet.[58] The CoE Convention on Cybercrime seeks to harmonise national laws relating to cybercrime, to improve domestic procedures for detecting, investigating and prosecuting such crimes and to provide arrangements for fast and reliable international cooperation on these matters. The CoE Convention on Cybercrime establishes a common minimum standard for domestic computer-related offences and provides for the criminalisation of nine

---

[55] ECtHR, decision of 12 January 2016 in case of Szabo and Vissy v. Hungary, Application No. 37138/14, at para. 68.

[56] ECtHR, decision of 12 January 2016 in case of Szabo and Vissy v. Hungary, Application No. 37138/14, at para. 73.

[57] ECtHR, decision of 12 January 2016 in case of Szabo and Vissy v. Hungary, Application No. 37138/14, at para. 73 referring to ECtHR, decision of 4 December 2015 in case of *Zakharov v. Russia*, Application No. 47143/06, at para. 249 and 266.

[58] CoE Convention on Cybercrime, ETS No.185, Budapest, 23 November 2001 (also known as Budapest Convention) which entered into force on 1 July 2004.

such offences,[59] including offences relating to unauthorised access to[60] and illicit tampering[61] with computer systems, programs or data, computer-related forgery[62] and fraud[63] and attempting, aiding or abetting the commission of such acts.[64]

The CoE Convention on Cybercrime also includes important procedural provisions which may facilitate investigations and gathering of evidence in connection with acts of terrorism involving use of the Internet.[65] These provisions apply to any criminal offence committed by means of a computer and the collection of evidence in electronic form and are subject to applicable safeguards provided for under domestic law.

It is apparent that the Parties to the CoE Convention on Cybercrime are subject to the resolutions by the United Nations Security Council and General Assembly which require the criminalisation of various forms of terrorism, facilitation of terrorism, support for terrorism and preparatory acts. In terrorism cases, the Parties often rely on offences that derive from those topic-specific treaties, as well as on additional offences in national legislation.

As an international treaty, the CoE Convention on Cybercrime is not focused specifically on terrorism. However, the substantive crimes in the CoE Convention on Cybercrime may be carried out as acts of terrorism, to facilitate terrorism, to support terrorism, including financially, or as preparatory acts.[66] In addition, the procedural and international mutual legal assistance tools in the Convention are available to terrorism and terrorism-related investigations and prosecutions.[67]

---

[59] See Art. 2–10 CoE Convention on Cybercrime.

[60] Art. 2 CoE Convention on Cybercrime.

[61] Art. 5 CoE Convention on Cybercrime.

[62] Art. 7 CoE Convention on Cybercrime.

[63] Art. 8 CoE Convention on Cybercrime.

[64] Art. 11 CoE Convention on Cybercrime.

[65] Art. 14–21 CoE Convention on Cybercrime.

[66] Section 2.3 of the T-CY Guidance Note #11, "Aspects of Terrorism covered by the Budapest Convention", adopted by the T-CY at its 16th Plenary, T-CY(2016)11, 15 November 2016.

[67] Sections 2.1 and 2.2 of the T-CY Guidance Note #11, "Aspects of Terrorism covered by the Budapest Convention", adopted by the T-CY at its 16th Plenary, T-CY(2016)11, 15 November 2016.

10.3.3 Procedural Provisions for Gathering Digital Evidence and Their Safeguards

The procedural powers for gathering digital evidence are enshrined in Articles 14–21 CoE Convention on Cybercrime. According to Article 14(2) CoE Convention on Cybercrime, these procedural powers may be used in specific criminal investigations or proceedings in any type of case. These specific procedural measures can be very useful in terrorism-related cases if a computer system was used to commit or facilitate the offence or if the evidence of that offence is stored in electronic form of if a suspect can be identified through subscriber information, including an IP-address. Therefore, the Parties to the CoE Convention on Cybercrime may use in terrorism cases expedited preservation of stored computer data[68] and of traffic data,[69] production orders,[70] search and seizure or stored computer data[71] as well as real-time collection of traffic data[72] and interception of content data.[73]

10.3.4 The Procedures for Gathering Digital Evidence

Article 16 CoE Convention on Cybercrime requests for LEAs and other competent national authorities the ability to order the expeditious preservation of specified computer data including traffic data. Such an order according to Article 16 CoE Convention on Cybercrime obliges the Internet service provider to save the data that were processed by this provider. The provider is not forced to start collecting data it would not normally store,[74] but the data which already exists has to be stored in a way that preserves its current quality and condition.[75] The provider is not obliged to transfer the relevant data to the requesting authority. Rather, Article 16 CoE Convention on Cybercrime authorises LEAs to prevent the deletion of the relevant data. The obligation to transfer data is regulated in Articles 17 and 18 CoE Convention on Cybercrime. By separating

[68] Article 16 CoE Convention on Cybercrime.
[69] Article 17 CoE Convention on Cybercrime.
[70] Article 18 CoE Convention on Cybercrime.
[71] Article 19 CoE Convention on Cybercrime.
[72] Article 20 CoE Convention on Cybercrime.
[73] Article 21 CoE Convention on Cybercrime.
[74] See No. 152 of the Explanatory Report to the CoE Convention on Cybercrime.
[75] See No. 159 of the Explanatory Report to the CoE Convention on Cybercrime.

the obligation to preserve the data from the obligation to disclose the data, the CoE Convention on Cybercrime offers the advantage of attaching different conditions to each obligation.

Article 17 CoE Convention on Cybercrime enables LEAs to order the expedited preservation and partial disclosure of traffic data which is extremely useful in cases requiring to trace back the route to a suspected individual and the need for immediate access to identify the path through which this communication was transmitted.

Based on Article 18 CoE Convention on Cybercrime, a provider can be obliged to disclose the data which it has preserved. The preservation of the data does not have to be based on a previous preservation order.[76] Rather, Article 18(a) CoE Convention on Cybercrime provides a general instrument for LEAs which is especially useful in cases not requiring access to hardware.[77] Article 18(b) CoE Convention on Cybercrime enables LEAs to order the submission of subscriber information which is an extremely useful tool in cases requiring IP-based investigations. If a LEA has identified an IP-address which was used in connection with an offence, this LEA needs to identify the person who used this IP-address at the time of the offence. Based on Article 18(1)(b) CoE Convention on Cybercrime, the provider is obliged to submit the subscriber information defined in Article 18(3) CoE Convention on Cybercrime.

Article 19 CoE Convention on Cybercrime introduces a data-related search and seizure procedure but does not specify the requirements which have to be met by investigators to carry out such investigations. Article 19 (1) CoE Convention on Cybercrime aims to establish an instrument that enables the search of computer systems which is as efficient as traditional procedures.[78] If the investigator of a LEA discovers during such a search that relevant information is stored on another computer system (e.g. cloud computing), Article 19 (2) CoE Convention on Cybercrime addresses the need to extend the search to that other system. Article 19 (3) CoE Convention on Cybercrime provides four important measures for receiving evidence which is acceptable in court proceedings: (a) an instrument to seizure a computer system, (b) an instrument to copy the data, (c) to

[76] Based on Article 16 CoE Convention on Cybercrime.

[77] Gercke, "Understanding Cybercrime: phenomena, challenges and legal response", ITU 2012, at 6.5.3, p. 248.

[78] This instrument has to be supplemented, see No. 187 of the Explanatory Report to the CoE Convention on Cybercrime.

maintain the integrity of copied data[79] and (d) an instrument that allows them to remove the data if it is illegal content or to ensure at least that this illegal content data can no longer be accessed. Finally, Article 19 (4) CoE Convention on Cybercrime enables the investigator of a LEA to compel a system administrator to assist LEAs because it is necessary for LEAs to identify the exact location of suspicious data. This provision not only obliges the system administrator to provide the necessary information to the investigator but also relieves the system administrator from contractual obligations or orders by his supervisors.[80] The scope of the obligation to support the investigator of a LEA extends only as far "as is reasonable", but the CoE Convention on Cybercrime does not define the term "reasonable". According to the Explanatory Report, "reasonable" may include disclosing a password or other security measure, but does not in general cover such disclosure where this would go along with unreasonable threats to the privacy of other users or data not included in the current search.[81]

Article 20 CoE Convention on Cybercrime introduces two different ways to collect traffic data. The term "traffic data" refers to data generated by computers during the communication process in order to route a communication from its origin to its destination. Therefore, "traffic data" includes IP-addresses identifying the partners of an Internet-related communication.[82] The first way of collecting such traffic data is according to Article 20(1)(a) CoE Convention on Cybercrime to impose an obligation on an Internet service provider to enable LEAs to collect the relevant data directly which generally requires the installation of an interface for LEAs to access the provider's infrastructure.[83] The second way enables LEAs to compel an Internet service provider according to Article 20(1)(b) CoE Convention on Cybercrime to collect data at their request allowing LEAs to benefit from the technical capacities and the knowledge of the provider. One of the major difficulties for investigations based on Article 20 CoE Convention on Cybercrime is the use of means of anonymous communication. Similarly, the use of public Internet terminals creates a comparable anonymity for its users, although the Court of Justice of the European

[79] See No. 197 of the Explanatory Report to the CoE Convention on Cybercrime.
[80] See No. 201 of the Explanatory Report to the CoE Convention on Cybercrime.
[81] See No. 202 of the Explanatory Report to the CoE Convention on Cybercrime.
[82] See No. 30 of the Explanatory Report to the CoE Convention on Cybercrime.
[83] See No. 220 of the Explanatory Report to the CoE Convention on Cybercrime.

Union has introduced a duty to identify users of a public WLAN to avoid liability for copyright and other offences committed using this WLAN.[84]

Article 21 CoE Convention on Cybercrime provides the possibility for LEAs to record data communications and to analyse the content if the LEAs already know who the communication partners are but have no information about the type of information exchanged. The content data affected by this provision includes files downloaded from websites or file-sharing systems, e-mails and chat/VoIP conversations. One of the most important difficulties for investigations based on Article 21 CoE Convention on Cybercrime is the use of encryption technology.[85]

10.3.5 The System of Safeguards

Because of the broad variety among existing national approaches of the Parties to the Convention on Cybercrime relating to safeguards and especially the way in which the Parties protect the rights of the suspect in the various criminal law systems, the CoE Convention on Cybercrime is designed to request its Parties to ensure that fundamental national and international standards of safeguards are applied. Article 15(1) CoE Convention on Cybercrime not only requests the application of already existing domestic conditions and safeguards also to Internet-related instruments but also defines the minimum standards by referring to fundamental frameworks such as the UDHR and the ICCPR of the United Nations as well as other international human rights instruments like the ECHR.

The reference in Article 15(1) CoE Convention on Cybercrime to the ECHR includes the protection of the right to respect for private and family life enshrined in Article 8 ECHR which appears most relevant for cybercrime investigations. To ensure the protection of privacy granted in Article 8 ECHR, the European Court of Human Rights (ECtHR) has developed a body of case law defining more precisely the standards that govern digital investigations and especially surveillance. This body of case law seems today one of the most important sources for international

[84] CJEU, decision of 15 September 2016 in case C-484/14, *Tobias Mc Fadden v. Sony Music Entertainment Germany GmbH*; Bisle/Frommer, CR 2017, pp. 54–63.

[85] Gercke, "Understanding Cybercrime: phenomena, challenges and legal response", ITU 2012, at 6.5.3, p. 259.

standards with respect to investigations related to communication.[86] The body of case law takes particularly into consideration the *gravity* of interference of the investigation,[87] the *purpose* of the interference of the investigation[88] and the *proportionality* of the interference of the investigation.[89] From the ECtHR's body of case law can be extracted the following four fundamental principles:

1. The need for a sufficient legal basis for investigation instruments.[90]
2. The requirement that the legal basis must be clear with regard to the rights of a suspect.[91]
3. The competences of LEAs need to be foreseeable.[92]
4. The surveillance of communication can only be justified in context of serious crime.[93]

In addition to these fundamental frameworks, Article 15(1) CoE Convention on Cybercrime expressly refers to the principle of proportionality which creates for Parties who are not Member States of the Council of Europe an obligation to develop the necessary safeguards.[94]

Article 15(2) CoE Convention on Cybercrime supplements these safeguards with an explicit reference to some of the most relevant safeguards

[86] Gercke, "Understanding Cybercrime: phenomena, challenges and legal response", ITU 2012, at 6.5.3, p. 245.

[87] ECtHR, decision of 12 April 1990 in case of *Kruslin v. France*, Application No. 11801/85, at para. 33.

[88] ECtHR, decision of 26 April 1985 in case of *Malone v. United Kingdom*, Application No. 8691/79, at para. 67.

[89] ECtHR, decision of 6 September 1978 in case of *Klass and others v. Germany*, Application No. 5029/71, at para. 42.

[90] ECtHR, decision of 12 April 1990 in case of *Kruslin v. France*, Application No. 11801/85, at para. 27.

[91] ECtHR, decision of 27 April 2004 in case of *Doerga v. The Netherlands*, Application No. 50210/99, at para. 50.

[92] ECtHR, decision of 12 April 1990 in case of *Kruslin v. France*, Application No. 11801/85, at para. 27 and ECtHR, decision of 26 April 1985 in case of *Malone v. United Kingdom*, Application No. 8691/79, at para. 67.

[93] ECtHR, decision of 6 September 1978 in case of *Klass and others v. Germany*, Application No. 5029/71, at para. 42.

[94] Gercke, "Understanding Cybercrime: phenomena, challenges and legal response", ITU 2012, at 6.5.3, p. 245.

including independent supervision, grounds justifying an application and the limitation of the scope and the duration of such power or procedure.[95]

All in all, the system of safeguards required by the CoE Convention on Cybercrime combines the ability of LEAs to use the instruments provided in Art. 14–21 CoE Convention on Cybercrime in a flexible way with the guarantee of effective safeguards and depends on the implementation of a graded system of safeguards. The decision which safeguard needs to be implemented with regard to which instrument is left to the national legislators of the Parties.[96] The ability to ensure an adequate protection of the rights of a suspected individual within a graded system of safeguards largely depends on how the potential impact of an investigation instrument is balanced with the related safeguards at national level.

10.3.6 Gathering Digital Evidence and International Cooperation

The CoE Convention on Cybercrime addresses the increasing importance of international cooperation in its Articles 23 to 35. Article 23 CoE Convention on Cybercrime defines the following three general principles for international cooperation in cybercrime investigations among Parties to the CoE Convention on Cybercrime:

- Extent: Parties are supposed to provide each other cooperation in international investigations to the widest extent possible.
- Scope: The general principles are applicable in any investigation involving the need to collect evidence in electronic form.
- Subsidiarity: The provisions of the CoE Convention on Cybercrime substitute neither provisions of international agreements pertaining to MLA and extradition nor relevant provisions of domestic law pertaining to international cooperation.

The drafters of the CoE Convention on Cybercrime emphasised that MLA should in general be carried out through the application of relevant treaties and similar arrangements for MLA. As a consequence, the CoE

[95] This list of most relevant safeguards in Article 15(2) CoE Convention on Cybercrime is not exclusive, see No. 146 of the Explanatory Report to the CoE Convention on Cybercrime.

[96] See No. 147 of the Explanatory Report to the CoE Convention on Cybercrime.

Convention on Cybercrime does not intend to create a separate general regime on MLA.[97]

The CoE Convention on Cybercrime requires parties to adopt a set of procedural powers to secure electronic evidence, such as search and seizure of computer systems,[98] production orders for data[99] and interception of communications.[100] These are subject to rule of law safeguards. They apply to electronic evidence in relation to any crime, including in relation to terrorist offences. International cooperation provisions also largely apply to cooperation in cases of electronic evidence, not just cybercrime.

According to Article 27(4) CoE Convention on Cybercrime, a Party to the Convention on Cybercrime may refuse a request for MLA if the request concerns an offence which the requested Party considers a political offence or an offence connected with a political offence. In an effort to overcome this obstacle for investigations in the context of terrorism, the CoE Convention on the Prevention of Terrorism[101] introduced in its Article 20 an exclusion of the political exception clause. The CoE Convention on the Prevention of Terrorism contains several terrorist-related offences such as public provocation to commit a terrorist offence[102] and recruitment[103] as well as training[104] for terrorism but does not contain specific provisions criminalising terrorism-related attacks against computer systems. Furthermore, the CoE Convention on the Prevention of Terrorism does not contain procedural instruments. Especially with regard to the investigation of Internet-related offences, specific procedural instruments are often required; e.g. identifying an offender who has incited terrorism using websites requires sophisticated instruments such as the expedited preservation of traffic data[105] and submission of subscriber information relating to web services.[106]

---

[97] Gercke, "Understanding Cybercrime: phenomena, challenges and legal response", ITU 2012, at 6.6.5, p. 274.

[98] Art. 19 CoE Convention on Cybercrime.

[99] Art. 16, 17 and 18 CoE Convention on Cybercrime.

[100] Art. 21 CoE Convention on Cybercrime.

[101] CoE Convention on the Prevention of Terrorism, CETS Nr. 196, Warsaw, 16 May 2005 which entered into force on 1 June 2007.

[102] Art. 5 CoE Convention on the Prevention of Terrorism.

[103] Art. 6 CoE Convention on the Prevention of Terrorism.

[104] Art. 7 CoE Convention on the Prevention of Terrorism.

[105] Art. 17 CoE Convention on Cybercrime.

[106] Art. 18 CoE Convention on Cyberdrime.

In practice, a core difficulty facing national LEAs is that electronic evidence needed is increasingly available only in foreign, sometimes unknown, multiple or shifting jurisdictions. MLA arrangements appear not always feasible or too cumbersome to secure volatile electronic evidence, although Article 25 CoE Convention on Cybercrime highlights the importance of fast communication and Article 26 CoE Convention on Cybercrime sets out regulations necessary for LEAs to inform foreign LEAs without jeopardising their own investigation. This is hardly surprising because both formal processes were developed to protect the integrity of a Party as a state as well as to safeguard the rights of the accused.[107] Bearing in mind the principle of national sovereignty, the procedural instruments provided by the CoE Convention on Cybercrime can only be used for investigations at the national level. If investigators realise that evidence has to be collected outside their national territory, they need to request MLA.

All but one procedural instrument established in the Articles 16–21 CoE Convention on Cybercrime[108] has a corresponding provision in the Articles 28–33 CoE Convention on Cybercrime enabling LEAs to apply the procedural instruments upon request of a foreign LEA. Only Article 18 CoE Convention on Cybercrime on production orders including on subscriber information has no corresponding provision in Chapter III on international cooperation of the CoE Convention on Cybercrime. As of 2015 therefore, the Cybercrime Convention Committee (T-CY) established a working group on criminal justice access to evidence stored in the cloud including through MLA (Cloud Evidence Working Group)[109] to identify solutions. The Recommendations of this Cloud Evidence Group were discussed by the plenary meeting of the Cybercrime Convention Committee[110] and by the international Octopus Conference in November 2016.[111]

There is full agreement that MLA must be made more efficient when it comes to electronic evidence. This includes training and allocation of

[107] Gercke, "Understanding Cybercrime: phenomena, challenges and legal response", ITU 2012, at 6.6.5, p. 276.

[108] Chapter III. on International co-operation of the CoE Convention on Cybercrime.

[109] Annex 4.2 of Cybercrime Convention Committee (T-CY), „Transborder access to data and jurisdiction: Options for further action by the T-CY", T-CY (2014)16, 3 December 2014.

[110] Cybercrime Convention Committee (T-CY), Meeting report, 16th Plenary, T-CY (2016)32, 14–15 November 2016.

[111] Octopus Conference "Cooperation against Cybercrime", 16–18 November 2016, Council of Europe, Strasbourg, France.

resources but also the establishment of emergency procedures, including, for example, in the case of terrorist threats.[112]

There is broad support for, but not yet full consensus on, a Guidance Note on the Production of Subscriber Information (Article 18 CoE Convention on Cybercrime).[113] Once adopted, this Guidance Note[114] would provide criminal justice authorities with the ability to request a service provider offering its service in the territory of a Party to produce subscriber information, for example, of a webmail or a social media account even if the data or the provider is in another Party's jurisdiction. This is already current practice, but the legal basis appears unclear. If the Guidance Note stands, Article 18 CoE Convention on Cybercrime could serve as the domestic legal basis.

There is also broad support for the preparation of an Additional Protocol to the CoE Convention on Cybercrime covering additional possibilities for mutual legal assistance, conditions for direct transborder access to data, provisions for direct cooperation with providers in other jurisdictions and provisions for the protection of personal data.[115] The Cybercrime Convention Committee may decide in June 2017 whether to go ahead with the negotiation of such Additional Protocol.

10.3.7 Gathering Digital Evidence Without MLA

The CoE Convention on Cybercrime provides in its Article 32 two scenarios in which a Party may have access to stored computer data without the authorisation of another Party:

[112] Agenda Item 7 of Cybercrime Convention Committee (T-CY), Meeting report, 16th Plenary, T-CY (2016)32, 14–15 November 2016; Cybercrime Convention Committee (T-CY), "T-CY assessment report: The mutual legal assistance provisions of the Budapest Convention on Cybercrime" T-CY(2013)17rev, 3 December 2014.

[113] Agenda Item 5 with regard to recommendation 2 of Cloud Evidence Group in Cybercrime Convention Committee (T-CY), Meeting report, 16th Plenary, T-CY (2016)32, 14–15 November 2016.

[114] T-CY Guidance Note #10 (DRAFT), "Production orders for subscriber information (Article 18 Budapest Convention)", Revised version as discussed by the T-CY at its 16th Plenary, T-CY(2015)16, 15 November 2016.

[115] Agenda Item 5 with regard to recommendation 5 of Cloud Evidence Group in Cybercrime Convention Committee (T-CY), Meeting report, 16th Plenary, T-CY (2016)32, 14–15 November 2016.

- The first scenario concerns access to publicly available (open-source) stored computer data regardless of where the data is located geographically.[116] An example of such publicly available data is information made available on websites without access control (such as passwords). If investigators were not allowed to access these websites, this could seriously hamper their investigation.
- The second scenario concerns access to data based on the consent of the person in control of this data.[117] When an investigator has obtained the lawful and voluntary consent of the person who has lawful authority to disclose the data, the investigator may access this data.

Whereas the first scenario appears widely accepted, the second scenario raises serious concerns because it probably contradicts fundamental principles of international law. Based on international law, investigators have to respect national sovereignty during an investigation.[118] Investigators are especially not allowed to carry out investigations in another state without the consent of the competent authorities in that state. The decision whether such permission should be granted is not in the hands of an individual but of the state authorities because interference with national sovereignty not only affects the rights of the individual but also state concerns. However, it may be argued that Parties ratifying the CoE Convention on Cybercrime partly waive their protection by the principle of national sovereignty and allow other countries to carry out investigations affecting their territory.[119] This argument is supported by the Guidance Note of the Cybercrime Convention Committee on the interpretation of Article 32(b) CoE Convention on Cybercrime which points out that this provision would not cover situations where the data are not stored in another Party or where it is uncertain where the data are located.[120]

Other concerns regarding the second scenario are that Article 32(b) CoE Convention on Cybercrime neither defines any procedures for the investigation nor safeguards the suspect's right to privacy, right to

---

[116] Art. 32(a) CoE Convention on Cybercrime.

[117] Art. 32(b) CoE Convention on Cybercrime.

[118] National sovereignty is a fundamental principle in international law, see: Roth (2005).

[119] Gercke, "Understanding Cybercrime: phenomena, challenges and legal response", ITU 2012, at 6.6.5, pp. 277–278.

[120] T-CY Guidance Notes, T-CY(2013)29rev, 8 December 2014, p. 22 on the notion of "transborder" and "location" regarding Transborder Access to data (Article 32).

protection of his personal data and procedural rights. The wording of this provision seems to suggest that not even limitations of national law are applicable which would apply to identical domestic investigations. However, the Guidance Note of the Cybercrime Convention Committee on the applicable law points out that access to data would not be permitted under Article 32(b) CoE Convention on Cybercrime if access or disclosure was not permitted domestically.[121] Further, it has to be pointed out that LEAs may only seek (!) permission of the person who has the lawful authority to disclose the data. This contrasts starkly with the instrument of a production order according to Article 18 CoE Convention on Cybercrime. Finally, it has to be pointed out that the standard hypothesis underlying Art. 32(b) CoE Convention on Cybercrime is that the person contacted to provide access to the data is physically located in the territory of the requesting Party leading the Cybercrime Convention Committee to suggest in its Guiding Note that LEAs should take into account that many Parties would object or consider it a criminal offence, if a person who is physically in their territory is directly approached by foreign LEAs seeking his or her cooperation.[122]

10.3.8 The 24/7 Network of Contacts

To increase the speed of international investigations, the CoE Convention on Cybercrime not only highlights in its Article 25 the importance of enabling the use of expedited means of communication but also obliges its Parties in its Article 35 to designate a contact point for MLA requests which is available without any time limitations. Especially to improve the efficiency of MLA requests, the Parties are obliged to establish these contact points and to ensure that they are able to carry out certain immediate action[123] as well as to maintain their service.[124]

[121] T-CY Guidance Notes, T-CY(2013)29rev, 8 December 2014, p. 22 on the applicable law regarding Transborder Access to data (Article 32).

[122] T-CY Guidance Notes, T-CY(2013)29rev, 8 December 2014, p. 23 on the location of the person consenting to provide access or disclose data regarding Transborder Access to data (Article 32).

[123] Article 35(1) CoE Convention on Cybercrime mentions as measures (a) technical advice, (b) preservation of data and (c) the collection of evidence, the provision of legal information and locating of suspects.

[124] Article 35(2) CoE Convention on Cybercrime requires for such a contact point (a) the capacity to carry out communications on an expedited basis and (b) the ability to co-ordinate with other national authorities on an expedited basis.

The CoE Convention on Cybercrime does not prescribe which national authority should be responsible for operating the national contact point of the 24/7 network. Nevertheless, the idea of the 24/7 network of contact points provides a useful answer to the challenges of fighting cybercrime associated especially with the speed of data exchange processes. Unfortunately, a study carried out in 2009 on the functioning of 24/7 points of contact in an international network fighting cybercrime[125] revealed that the full potential of such a network is not (yet) used because not all Parties of the CoE Convention on Cybercrime had created a functioning contact point and not all contact points were used to their full capacity or known domestically.

10.4 Regional Legal Framework: European Union

In their fight against terrorism, Member States of the European Union have to respect human rights and the rule of law not only because of their obligations under the ECHR and the UDHR but also because of the Charter of Fundamental Rights of the European Union[126] (Charter of Fundamental Rights).

10.4.1 Charter of Fundamental Rights

The European Union (EU) has established the Charter of Fundamental Rights of the European Union (2000) for the protection of human rights. Concerning the protection of privacy, the Charter of Fundamental Rights comprises not only the right to respect for private and family life[127] but also the right to protection of personal data[128] implying a more coherent approach. The guarantees of the Charter of Fundamental Rights also include the freedom of expression and information,[129] freedom of

[125] CoE Economic Crime Division, „The functioning of 24/7 points of contact for cybercrime“, Discussion Paper, 2 April 2009.

[126] Charter of Fundamental Rights of the European Union, Official Journal of the European Communities, 2000/C 364/01, 18 December 2000.

[127] Art. 7 Charter of Fundamental Rights.

[128] Art. 8 Charter of Fundamental Rights.

[129] Art. 10 Charter of Fundamental Rights.

assembly and of association[130] as well as the right to a fair trial[131] and the presumption of innocence.[132]

10.4.1.1 *Applicability to Secret Electronic Surveillance*

Article 51 Charter of Fundamental Rights demands the Member States of the EU to respect the rights and to observe the principles laid down in the Charter of Fundamental Rights only when they are implementing Union law. Member States are implementing Union law when national legislation falls within the scope of European Union law which automatically opens the jurisdiction of the Court of Justice of the European Union (CJEU) to guide the interpretation of the Charter of Fundamental Rights so that national courts can determine whether a national legislation is compatible with the fundamental rights enshrined in the Charter of Fundamental Rights.[133]

In case LEAs use an automated tool in secret investigations for searching the Internet and the dark web for evidence, the protection of privacy and personal data is crucial for individuals whose activities and connections are examined.

10.4.1.2 *Exemption from General Data Protection Regulation*

Most relevant for the guarantees of the right to respect for private and family life[134] and the right to protection of personal data[135] is Article 2(2)(a)-(d) General Data Protection Regulation[136] which provides an exemption for LEAs from the scope of the General Data Protection Regulation and causes surveillance to fall under Union law (previously Article 13 of the Data Protection Directive[137]).

---

[130] Art. 11 Charter of Fundamental Rights.

[131] Art. 47 Charter of Fundamental Rights.

[132] Art. 48 Charter of Fundamental Rights.

[133] CJEU, decision of 26 February 2013 in case *Akerberg Fransson*, C-617/10, at para. 19.

[134] Art. 7 Charter of Fundamental Rights.

[135] Art. 8 Charter of Fundamental Rights.

[136] Regulation (EU) 2016/679 of the European Parliament and of the Council of 27 April 2016 on the protection of natural persons with regard to the processing of personal data and on the free movement of such data, and repealing Directive 95/46/EC (General Data Protection Regulation), Official Journal L 119/1, 4 May 2016.

[137] Directive 95/46/EC of the European Parliament and of the Council of 24 October 1995 on the protection of individuals with regard to the processing of personal data and on the free movement of such data, Official Journal L 281, 23/11/1995, p. 31.

10.4.1.3 Exemptions in ePrivacy Directive and Proposed Regulation on Privacy and Electronic Communication

The ePrivacy Directive ensures the protection of the fundamental right to respect for private and family life, the confidentiality of communications and the protection of personal data in the electronic communications sector. It also guarantees the free movement of electronic communications data, equipment and services in the Union. It implements in the Union's secondary law the fundamental right to the respect for private life, with regard to communications, as enshrined in Article 7 Charter of Fundamental Rights. According to its Article 1(3), the ePrivacy Directive shall not apply in any case to activities concerning (among others) public security and the activities of the State in areas of criminal law. This exemption for LEAs from the scope of the ePrivacy Directive also causes surveillance to fall under Union law.

Because consumers and businesses increasingly rely on Internet-based services enabling interpersonal communications such as Voice over IP, instant messaging and web-based e-mail services, instead of traditional communications services, the European Commission has proposed an ePrivacy Regulation[138] on 10 January 2017. These over-the-top communications services ("OTTs") are in general not subject to the current Union electronic communications framework, including the ePrivacy Directive. However, also this proposed ePrivacy Regulation is not to apply to activities of LEAs "for the purposes of the prevention, investigation, detection or prosecution of criminal offences" according to Art. 2(2)(d) ePrivacy Regulation. This exemption of activities of LEAs from the scope of the proposed ePrivacy Regulation will perpetuate that surveillance falls under Union law.

10.4.1.4 Restrictions on Freedom Provided in TFEU

The applicability of the Charter of Fundamental Freedoms also results from the fact that acts of surveillance may affect the prohibition of restrictions on the freedom to provide services within the EU in Article 56 of Treaty of Functioning of the European Union (TFEU).[139] According to

[138] European Commission, Proposal for a Regulation of the European Parliament and of the Council concerning the respect for private life and the protection of personal data in electronic communications and repealing Directive 2002/58/EC (Regulation on Privacy and Electronic Communications), COM(2017) 10 final, 10 January 2017.

[139] CJEU, decision of 26 February 2013 in case *Akerberg Fransson*, C-617/10, at para. 29.

the CJEU, even in a situation where action of a Member States is only partially determined by EU law, the "implementation" requirement of Article 51(1) Charter of Fundamental Rights is met whenever a national court is called upon to review whether fundamental rights are complied with by a national provision or measure.[140]

10.4.1.5 Lines of Case Law Synchronising Privacy Protection under Charter of Fundamental Rights and under ECHR

In the area of privacy and data protection, the CJEU has developed a line of case law which expounds Articles 7 and 7 Charter of Fundamental Freedoms in combination with Article 8 ECHR and refers to the established line of case law by the ECtHR on the guarantee of privacy under the ECHR.[141] In this context, it has to be pointed out that also the ECtHR refers in its more recent case law to the principles developed by the CJEU regarding the interpretation of Articles 7 and 8 Charter of Fundamental Rights.[142] This kind of cross-referencing to each other's line of case law appears to be a rather recent phenomenon but allows, nevertheless, to expect a uniform interpretation of the protection of privacy in the future.[143]

10.4.1.6 Minimum Standards for Privacy Protection

Furthermore, the CJEU expressly mentions "minimum safeguards" for individuals against the risk of abuse and unlawful access of data retained by LEAs in their fight against terrorism.[144] With these "minimum safeguards", the CJEU refers to the line of case law of the ECtHR described at Sect. 10.2.1 above establishing coherent minimum standards for national surveillance measures without formulating its own detailed catalogue of minimum requirements. This reference to the ECtHR's line of case law leads to the conclusion that the cumulative minimum standards established by the ECtHR are to be applied under the Charter of

---

[140] CJEU, decision of 26 February 2013 in case *Akerberg Fransson*, C-617/10, at para. 29.

[141] CJEU, decision of 21 December 2016 in case *Tele2 Sverige AB*, C-203/15 and C-698/15, at paras. 119 and 120; CJEU, decision of 8 April 2014 in case *Digital Rights Ireland*, C-293/12 and C-592/12, at para. 35, 47, 54.

[142] ECtHR, decision of 12 January 2016 in case of *Szabo and Vissy v. Hungary*, Application No. 37138/14, at para. 68, 70, 73: ECtHR, decision of 4 December 2015 in case of *Zakharov v. Russia*, Application No. 47143/06, at para. 147.

[143] Boehm/Andrees, CR 2016, pp. 146–154.

[144] CJEU, decision of 8 April 2014 in case *Digital Rights Ireland*, C-293/12 and C-592/12, at para. 54.

Fundamental Rights as well. Indeed, the CJEU goes on to examine each of the exact criteria developed by the ECtHR:

- Restrictions of individuals affected by the surveillance measure[145]
- Access restrictions to collected data to ensure their availability for serious crimes only[146]
- Limitation of data retention period[147]
- Guarantee of data security[148]

According to the CJEU, the retention of surveillance data requires an explicit reason for the collection of the data[149] and creates a need for a threat to public security causing the collection of data.[150]
As a result, the protection of the right to respect for private and family life[151] and of the right to protection of personal data[152] under the Charter of Fundamental Rights appears currently fully synchronised with the protection of the right to respect for private and family life[153] under the ECHR.

10.4.2 *Directive on Combating Terrorism*

The ratification of the Lisbon Treaty introduced a comprehensive mandate of the EU for cybercrime legislation as of 2009. Most relevant with regard to cybercrime is Article 83 TFEU.[154] It enables the EU to establish minimum rules concerning the definition of criminal offences and

---

[145] CJEU, decision of 8 April 2014 in case *Digital Rights Ireland*, C-293/12 and C-592/12, at para. 58.
[146] CJEU, decision of 8 April 2014 in case *Digital Rights Ireland*, C-293/12 and C-592/12, at para. 60.
[147] CJEU, decision of 8 April 2014 in case *Digital Rights Ireland*, C-293/12 and C-592/12, at para. 63.
[148] CJEU, decision of 8 April 2014 in case *Digital Rights Ireland*, C-293/12 and C-592/12, at para. 66.
[149] CJEU, decision of 8 April 2014 in case *Digital Rights Ireland*, C-293/12 and C-592/12, at para. 58.
[150] CJEU, decision of 8 April 2014 in case *Digital Rights Ireland*, C-293/12 and C-592/12, at para. 59.
[151] Art. 7 Charter of Fundamental Rights.
[152] Art. 8 Charter of Fundamental Rights.
[153] Art. 8 ECHR.
[154] Regarding the impact of the reform on the harmonization of criminal law see: Peers (2008), pp. 507 *et seq.*; Zeder (2008), pp. 209 et seq; Gercke (2010), pp. 75 *et seq.*

sanctions in relation to serious crime with a cross-border dimension. Article 83(1) TFEU lists terrorism and computer crime and organised crime as key areas among others.

Based on this mandate and the European Commission presented, the proposal for a Directive on Combating Terrorism[155] in the aftermath of the Paris attacks on 13 November 2015. This proposed Directive on Combating Terrorism takes into account the requirements stemming from the United Nations Security Council Resolution 2178 (2014) and the Additional Protocol to the CoE Convention on the Prevention of Terrorism as well as from the standards of the Financial Action Task Force (FATF)[156] regarding the financing of terrorism.

In March 2016, the Council of the EU presented its general approach[157] suggesting a compromise text of the proposed Directive on Combating Terrorism. On 30 November 2016, the Permanent Representatives Committee (Coreper) confirmed the agreement reached by the Slovak presidency with the European Parliament on the Directive on Combating Terrorism which was then also confirmed by the European Parliament's Committee on Civil Liberties, Justice and Home Affairs on 5 December 2016. This mutual confirmation paved the way for the final adoption of the Directive on Combating Terrorism on 15 March 2017.[158]

10.4.2.1 Harmonised Framework for Criminalising Terrorist-Related Activity

To respond to the evolving terrorist threat, the Directive on Combating Terrorism strengthens the EU's legal framework in preventing terrorist attacks by criminalising acts such as receiving training for terrorism[159] and travel for terrorist purposes,[160] as well as organising or facilitating such

[155] European Commission, Proposal for a Directive of the European Parliament and of the Council on Combating Terrorism and Replacing Council Framework Decision 2002/475/JHA on Combating Terrorism, 2 December 2015.

[156] FATF (2012).

[157] Council of the European Union, Proposal for a Directive of the European Parliament and of the Council on Combating Terrorism and Replacing Council Framework Decision 2002/475/JHA on Combating Terrorism – General Approach, 3 March 2016.

[158] Directive (EU) 2017/541 of the European Parliament and of the Council of 15 March 2017 on Combating Terrorism and Replacing Council Framework Decision 2002/475/JHA and amending Council Decision 2005/671/JHA, Official Journal L 88/6, 31 March 2017.

[159] Article 8 Directive on Combating Terrorism.

[160] Article 9 Directive on Combating Terrorism.

travel.[161] The Directive on Combating Terrorism also reinforces the rights for the victims of terrorism.[162] The Directive on Combating Terrorism strengthens and updates the existing Framework Decision 2002/475/JHA, in particular,[163] because it criminalises:

- *Travelling for terrorist purposes,*[164] to counter especially the phenomenon of foreign terrorist fighters. The compromise reached between the institutions in the trialogue will ensure that, e.g. travel to conflict zones with the purpose to join the activities of a terrorist group or travel to a EU Member State with the purpose to commit a terrorist attack will be made punishable.
- The *organisation and facilitation of such travels,*[165] including through logistical and material support (e.g. the purchase of tickets or planning itineraries)
- *Providing or receiving training for terrorist purposes,*[166] e.g. in the making or use of explosives, firearms, noxious or hazardous substances mirroring the already existing provision of knowingly providing such training;

Providing or collecting funds[167] with the intention or the knowledge that they are to be used to commit terrorist offences and offences related to terrorist groups or terrorist activities

10.4.2.2 Protection of Victims of Terrorism

The Directive on Combating Terrorism also complements the legislation on *rights for victims of terrorism.* In this respect, the Directive on Combating Terrorism not only adjusts for the terrorist offences the definition of "victims" provided in Article 2 Directive 2012/29/EU[168] but also

---

[161] Article 10 Directive on Combating Terrorism.

[162] Articles 24, 25 and 26 Directive on Combating Terrorism.

[163] The bullet points presented here are based on the press release 716/16 by Council of EU, "Directive on combatting terrorism: Council confirms agreement with Parliament", 5 December 2016.

[164] Article 9 Directive on Combating Terrorism.

[165] Article 10 Directive on Combating Terrorism.

[166] Article 7 and 8 Directive on Combating Terrorism.

[167] Article 11 Directive on Combating Terrorism.

[168] Recital 27 Directive on Combating Terrorism adjusting for terrorist offences the definition of "victim" in Article 2 of Directive 2012/29/EU of the European Parliament and of the Council of 25 October 2012 establishing minimum standards on the rights, support and protection of victims of crime, and replacing Council Framework Decision 2001/220/JHA, Official Journal of the EU, L 315/57, 14 November 2012.

includes a catalogue of services to meet the specific needs of victims of terrorism,[169] such as the right to receive immediate access to professional support services providing medical and psychosocial treatments[170] and for as long as necessary thereafter taking into account that specific needs of victims of terrorism may evolve in time,[171] or the right to receive legal or practical advices[172] as well as assistance with compensation claims.[173] Further, the Directive on Combating Terrorism also strengthens the emergency response mechanisms immediately after an attack.[174]

10.4.2.3 Investigative Measures and Tools

According to Article 21 Directive on Combating Terrorism,[175] Member States have to take the necessary measures to promptly remove or to block online content publicly inciting to commit terrorist offences requiring such measures to be transparent, necessary and proportionate. Further, the Directive on Combating Terrorism introduces enhanced rules for exchange of information between Member States related to terrorist offences gathered in criminal proceedings.[176]

Most relevant for the use of a sophisticated automated tool, the Directive on Combating Terrorism addresses the use of investigative tools by LEAs. According to Article 20(1) Directive on Combating Terrorism, Member States shall take the necessary measures to ensure that "effective investigative tools" are available to those responsible for investigating and prosecuting the offences referred to in Articles 3–12 Directive on Combating Terrorism. As examples for such "effective investigative tools", the provision mentions those measures used in organised crime and other serious crime cases. While requiring such investigative tools to take into account the principle of proportionality and the nature and seriousness of the offences under investigation in accordance with national law, these investigative tools should according to Recital 21 Directive on Combating Terrorism include the following:

---

[169] Article 24(3) Directive on Combating Terrorism.

[170] Article 24(3)(a) and Recital 29 Directive on Combating Terrorism.

[171] Recital 29 Directive on Combating Terrorism.

[172] Article 22(3)(b) Directive on Combating Terrorism.

[173] Article 22(3)(c) and Recital 28 Directive on Combating Terrorism.

[174] Article 24(2) and (4) and Recital 29 Directive on Combating Terrorism.

[175] Read together with Recitals (22) and (23) Directive on Combating Terrorism.

[176] Recital 25 Directive on Combating Terrorism.

- "The interception of communications"
- The "covert surveillance including electronic surveillance"
- The taking and the fixing of "audio recordings in private or public vehicles and places"
- The taking and the fixing of "visual images of persons in public vehicles and places"
- "Financial investigations"

Especially the inclusion of investigative tools for "covert surveillance including electronic surveillance" paves the way for the use of an automated tool in police investigations of the surface web and the dark web. According to the second sentence of Recital 21 Directive on Combating Terrorism, the right to the protection of personal data should be respected when such an effective investigative tool is used.

10.4.2.4 Multiple Jurisdictions

In this context, it is important that the Directive on Combating Terrorism addresses the question of multiple jurisdiction only for the prosecution of an offender in its Article 19(3). This provision states in its first sentence that the Member States concerned shall "cooperate in order to decide which of them will prosecute the offender with the aim, if possible, of centralising the proceedings in a single Member State". Article 19(3) sentence 3 lit. a) – d) Directive on Combating Terrorism provides a list of four factors to be taken into account. Unfortunately, a comparable provision for the investigation of the Directive's terrorist offences is not included. Merely, Article 19(1) sentence 1 lit. a) – e) Directive on Combating Terrorism introduces five criteria based on which each Member State shall take the necessary measures to establish its jurisdiction over the Directive's offences, and Article 19(1) sentence 2 Directive on Combating Terrorism explicitly mentions that each Member State "may extend its jurisdiction if the offence is committed in the territory of another Member State". However, this invitation to extend their jurisdiction does not affect, let alone alter, for Member States of the EU the system for gathering digital evidence established by the CoE Convention on Cybercrime.

10.4.2.5 Observation of Fundamental Rights and Freedoms

Recital 35 Directive on Combating Terrorism[177] explicitly requires Member States to implement the Directive into national law in accordance with the rights set out in Titles II, III, V and VI of the Charter of Fundamental Rights[178] and the freedom of movement as set forth in Article 21(1) TFEU and Directive 2004/38/EC.[179] According to this Recital, Member States also have to take into account the ECHR, the ICCPR and other human rights obligations under international law.

Finally, it seems noteworthy to observe for the exchange of information when developing a sophisticated automated tool for use by LEAs that nothing in the Directive on Combating Terrorism should be interpreted as being intended to reduce or restrict the dissemination of information for scientific, academic or reporting purposes.[180]

10.4.2.6 Resulting Framework for Use of Automated Tools

The Directive on Combating Terrorism was to be transposed into the national law of Member States by 8 September 2018[181] and is particularly relevant for the development and use of automated tools because, according to Art. 20(1) Directive on Combating Terrorism, Member States shall take necessary measures to ensure that effective investigative tools are available to Law Enforcement Agencies (LEAs) investigating or prosecuting the offences related to terrorism defined in Art. 3 and Art. 4 Directive on Combating Terrorism.[182] These offences concern:

- Public provocation to commit a terrorist offence, Art. 5 Directive on Combating Terrorism

---

[177] See also the last sentence of Recital 22 Directive on Combating Terrorism.

[178] This includes the freedom of expression and information (Article 11), the right to respect for private and family life (Article 7) and the right to protection of personal data (Article 8) as well as the presumption of innocence (Article 48).

[179] Directive 2004/38/EC of the European Parliament and of the Council of 29 April 2004 on the right of citisens of the Union and their family members to move and reside freely within the territory of the Member States amending Regulation (EEC) No 1612/68 and repealing Directives 64/221/EEC, 68/360/EEC, 72/194/EEC, 73/148/EEC, 75/34/EEC, 75/35/EEC, 90/364/EEC, 90/365/EEC and 93/96/EEC, Official Journal of the EU, L 158/77, 30 April 2004.

[180] Recital 40 Directive on Combating Terrorism.

[181] Art. 28(1) Directive on Combating Terrorism.

[182] Definition of terrorist offences in Art. 3 and of offences related to terrorist groups in Art. 4 Directive on Combating Terrorism.

- Recruitment for terrorism, Art. 6 Directive on Combating Terrorism
- Providing training for terrorism, Art. 7 Directive on Combating Terrorism
- Receiving training for terrorism, Art. 8 Directive on Combating Terrorism
- Travelling for the purpose of terrorism, Art. 9 Directive on Combating Terrorism
- Organising or otherwise facilitating travelling for the purpose of terrorism, Art. 10 Directive on Combating Terrorism
- Terrorist financing, Art. 11 Directive on Combating Terrorism and
- Other offences related to terrorist activities, Art. 12 Directive on Combating Terrorism

This establishes a harmonised framework for criminalising terrorist-related activity in the EU overall and accordingly the investigation and prosecution of such offences. At the same time, the Directive on Combating Terrorism also reinforces the rights for the victims of terrorism in Articles 24, 25 and 26 Directive on Combating Terrorism.

As examples for such "effective investigative tools", Art. 20(1) Directive on Combating Terrorism refers to "those tools which are used in organised crime or other serious crime cases".

According to Recital 21 of the Directive on Combating Terrorism, the use of such investigative tools has to not only be in accordance with national law and targeted but also take into account the principle of proportionality and the nature and seriousness of the offences under investigation as well as respect the right to the protection of personal data. Such investigative tools should, where appropriate, include, for example:

- "The interception of communications"
- "Covert surveillance including electronic surveillance"
- "The taking and the keeping of audio recordings" (in private or public vehicles and places) "and of visual images of persons" (in public vehicles and places) and "financial investigations"[183]

Especially the inclusion of investigative tools for "covert surveillance including electronic surveillance" paves the way for the development and use of an automated tool.

[183] See last sentence of Recital 21 of the Directive on Combating Terrorism.

If the (repressive) prosecution of an offence falls within the jurisdiction of more than one Member State, then Art. 19(3) sentence 3 lit. a) – d) Directive on Combating Terrorism provides a list of four factors the Member States concerned have to consider in order to decide which of them will prosecute the offender aiming to centralise the proceedings in a single Member State.
Unfortunately, a comparable provision for the (progressive) investigation of the Directive's terrorist offences is not included. Although Article 19(1) Sentence 2 Directive on Combating Terrorism allows each Member State to extend its jurisdiction to national terrorist-related offences listed in Sentence 1 but committed in the territory of a Member State, this neither affects nor alters for Member States of the EU the system for gathering digital evidence established by the CoE Convention on Cybercrime.

10.4.3 Directive (EU) 2016/680 for Data Protection in the Police and Criminal Justice Sectors

On 5 May 2016, Directive (EU) 2016/680 on the Protection of Natural Persons with regard to the Processing of Personal Data by Competent Authorities for the Purposes of the Prevention, Investigation, Detection or Prosecution of Criminal Offences or the Execution of Criminal Penalties and on the Free Movement of Such Data[184] entered into force, and Member States have to transpose it into their national law by 6 May 2018.

10.4.3.1 Legislative Competence of the European Union

Directive (EU) 2016/680 has been adopted in order to ensure a high level of data protection while improving cooperation in the fight against terrorism and other serious crime. After the Treaty of Lisbon came into effect, the protection of natural persons in relation to the processing of personal data is expressly recognised as a fundamental right. Article 8(1) Charter of Fundamental Rights and Article 16(1) TFEU provide that everyone has the right to the protection of personal data concerning him or her. However, Declaration 21, annexed to the final act of the intergovernmental conference which adopted the Treaty of Lisbon, acknowledges that the specific nature of the security field merits special legislative

[184] Directive (EU) 2016/680 of the European Parliament and of the Council of 27 April 2016 on the protection of natural persons with regard to the processing of personal data by competent authorities for the purposes of the prevention, investigation, detection or prosecution of criminal offences or the execution of criminal penalties, and on the free movement of such data, and repealing Council Framework Decision 2008/977/JHA.

treatment. According to the European institutions' approach, processing in the police and criminal justice context should be differentiated from all other personal data processing. The protection and free movement of data processed by competent authorities for the purposes of the prevention, investigation, detection or prosecution of criminal offences or the execution of criminal penalties has been regulated by a directive, allowing Member States a certain level of flexibility while incorporating it into their respective national laws.

10.4.3.2 Brief Overview of Contents

Directive (EU) 2016/680 aims at balancing the data protection objectives with the security policy objectives, and while certainly contributing to the creation of a less fragmented general framework, it does not solve all the shortcomings which had emerged before its adoption. Directive (EU) 2016/680 comprises ten chapters which can be divided into two parts:

The *first part* of Directive (EU) 2016/680 consists of Chapters I–V which describes the following:

- The scope[185]
- The general principles relating to processing of personal data[186]
- The rights of the data subject[187]
- The obligations of the controllers and the processors,[188] the technical and organisational measures to ensure security of personal data, which have to be adopted by them,[189] as well as the designation of a data protection officer[190]
- The regulation of transfer of personal data to third countries or international organisations[191]

The *second part* of Directive (EU) 2016/680 regulates:

[185] Articles 1–3 Directive (EU) 2016/680, chapter I.
[186] Articles 4 – 11 Directive (EU) 2016/680, chapter II.
[187] Articles 12 – 18 Directive (EU) 2016/680, chapter III.
[188] Articles 19 – 28 Directive (EU) 2016/680, chapter IV, section 1.
[189] Articles 29 – 31 Directive (EU) 2016/680, chapter IV, section 2.
[190] Articles 32 – 34 Directive (EU) 2016/680, chapter IV, section 3.
[191] Articles 35 – 40 Directive (EU) 2016/680, chapter V.

- The independent status,[192] the competence, tasks and powers[193] of the independent supervisory authorities and establishes the right to lodge a complaint with a supervisory authority
- The cooperation between Member States by mutual assistance[194]

The right to an effective judicial remedy against a controller or processor and the right to compensation for any person who has suffered material or non-material damage as a result of an unlawful processing of personal data[195]

10.4.3.3 Scope

Directive (EU) 2016/680 applies to the processing of personal data by competent authorities "for the purposes of the prevention, investigation, detection or prosecution of criminal offences or the execution of criminal penalties", Article 2(1) Directive (EU) 2016/680 in connection with Art. 1(1) Directive (EU) 2016/680. The use of an automated TENSOR tool in a criminal investigation falls clearly within the scope of Directive (EU) 2016/680.

10.4.3.4 Data Processing in the Course of Criminal Investigations

It is interesting to note that Recital 49 Directive (EU) 2016/680 seems to suggest that where personal data are processed in the course of "a criminal investigation", Member States may provide for the exercise of the right to information,[196] access[197] and rectification or erasure[198] of personal data to be carried out in accordance with their national law. Read together with Article 18 as well as Recitals 20 and 107 Directive (EU) 2016/680, this appears to provide an opening for different national laws under the framework of Directive (EU) 2016/680. Because of this ambiguity, the real added value of Directive (EU) 2016/680 will depend on its implementation in national law and the willingness of national courts to ensure that Directive (EU) 2016/680 is applied in a uniform manner across the EU.

---

[192] Articles 41–44 Directive (EU) 2016/680, chapter VI, section 1.

[193] Articles 45 – 49 Directive (EU) 2016/680, chapter VI, section 2.

[194] Articles 50 – 51 Directive (EU) 2016/680, chapter VII.

[195] Articles 52 – 57 Directive (EU) 2016/680, chapter VIII. The final two of Directive (EU) 2016/680 are about implementing acts, chapter IX, and final provisions, chapter X.

[196] Article 13 Directive (EU) 2016/680.

[197] Article 14 Directive (EU) 2016/680.

[198] Article 16 Directive (EU) 2016/680.

10.4.3.5 Data Processing Outside the Scope of Union Law

Directive (EU) 2016/680 does not regulate the processing of data in the course of an activity which falls outside the scope of Union law, Article 2(3)(a) Directive (EU) 2016/680. Recital 14 Directive (EU) 2016/680 suggests to interpret Article 2(3)(a) Directive (EU) 2016/680 as relating to activities concerning national security, activities of agencies or units dealing with national security issues and the processing of personal data by the Member States when carrying out activities which fall within the scope of Chapter 2 of Title V of the TEU. As consequence, the wording of Article 2(3) Directive (EU) 2016/680 appears to be in conflict with the inclusion of "safeguarding against and the prevention of threats to public security" in Article 1(1) Directive (EU) 2016/680. The concept of activities concerning national security is not defined in Directive (EU) 2016/680, but seems to include "activities of safeguarding against and prevention of threats to public security". Until the CJEU will guide the interpretation of this contradiction, the scope of Directive (EU) 2016/680 will depend on the interpretation that national courts will give to the expression "activity which falls outside the scope of Union law" and of the way the Member States decide to implement Directive (EU) 2016/680.

10.4.3.6 Data Processing by EU Institutions, Bodies, Offices and Agencies

Finally, Directive (EU) 2016/680 does not apply to the processing of personal data by the Union institutions, bodies, offices and agencies, Article 2(3)(b) Directive (EU) 2016/680. The data processing by the European institutions and bodies will continue to be governed by Regulation (EC) No. 45/2001. However, the European Commission has presented a Proposal for a Regulation on the protection of individuals with regard to the processing of personal data by the Union institutions, bodies, offices and agencies and on the free movement which will repeal Regulation (EC) No. 45/2001 on 10 January 2017.[199] This proposed Regulation repealing Regulation (EC) No. 45/2001 aims to bring the

---

[199] European Commission, Proposal for a Regulation of the European Parliament and of the Council on the protection of individuals with regard to the processing of personal data by the Union institutions, bodies, offices and agencies and on the free movement of such data, and repealing Regulation (EC) No 45/2001 and Decision No 1247/2002/EC, COM(2017) 8 final, 10 January 2017.

level of data protection at EU institutions, bodies, offices and agencies in line not only with the GDPR but also with Directive (EU) 2016/680.[200]

10.4.3.7 Minimum Harmonisation Within the EU

Unlike the 2008 Framework Decision, Directive (EU) 2016/680 regulates the processing of personal data by Member States and not only intra-Member States exchanges of data. Nevertheless, Directive (EU) 2016/680 is still far from ensuring maximum harmonisation of data processing in the criminal field. Article 1(3) Directive (EU) 2016/680 states that Directive (EU) 2016/680 shall not preclude Member States from providing higher safeguards than those established in Directive (EU) 2016/680 for the protection of the rights and freedoms of the data subject. As a result, Directive (EU) 2016/680 introduces only a minimum harmonisation.

10.4.3.8 Comparison of Principles for Data Processing with GDPR

Several principles relating to processing of personal data are the same as those enshrined in the GDPR. However, because of the peculiarity of the field, while the basic data protection principles are included in its text, some of those set out in the GDPR are not included in Directive (EU) 2016/680. For example,

as far as the *characteristics the data* should have in order to be processed by the competent authorities are concerned, it may be observed that not all the conditions required by the GDPR in order to consider the data processing lawful and fair need to be met under Directive (EU) 2016/680.

The *consent of the data subject* is not a necessary condition for processing personal data by the competent authorities according to Recital 35 Directive (EU) 2016/680 when they order natural persons to comply with requests made in order to perform the tasks of preventing, investigating, detecting or prosecuting criminal offences. Where the data subject is required to comply with a legal obligation, the data subject has no genuine and free choice, so that the reaction of the data subject could not be considered to be a freely given indication of his or her wishes. Whether the correct balance between individual data protection and the interests of the

---

[200] Recitals 9 and 10 of the Proposal for a Regulation of the European Parliament and of the Council on the protection of individuals with regard to the processing of personal data by the Union institutions, bodies, offices and agencies and on the free movement of such data, and repealing Regulation (EC) No 45/2001 and Decision No 1247/2002/EC, COM(2017) 8 final, 10 January 2017.

police and criminal justice process is respected depends once again on how Member States will implement the exemptions contained in Directive (EU) 2016/680.

Directive (EU) 2016/680 also allows Member States to adopt legislative measures restricting the data subject's rights to information,[201] access[202] and rectification[203] in an attempt to strike a balance between the individual right to data protection and the processing interests and concerns of the police and other LEAs. If exercised to their fullest extent, these rights would undermine much of the work done by the police and the competent authorities within the criminal justice system. The level of flexibility accorded to this end depends once more on the breadth of national legislative measures implementing Directive (EU) 2016/680, which can restrict, wholly or partly, the data subject's right in order to assure the due performance of investigations and protect national security, as set out in Article 15 Directive (EU) 2016/680.

10.4.3.9 Independent Supervisory Authority

The final important element of the EU data protection model refers to the establishment of an independent supervisory authority entrusted with the task of monitoring the application of data protection law within the respective Member State. Directive (EU) 2016/680 permits assignment of this role to the authority established for similar purposes under the GDPR. Data Protection Authorities, as independent supervisory authorities, have been already introduced by the Data Protection Directive and have become the basic mechanism for enforcement and monitoring of data protection in the EU.

An ostensibly significant change brought by the EU data protection reform package to the EU data protection systems concerns the replacement of the old Article 29 Data Protection Working Party by the European Data Protection Board. The European Data Protection Board will replace the Article 29 Working Party but, as far as Directive (EU) 2016/680 is concerned, apparently only because it will essentially retain the same powers. In this respect, it should be noted that, while the GDPR assigns a central role to the European Data Protection Board (especially in the consistency mechanism), no such role is provided for in Directive (EU)

[201] Article 13(3) Directive (EU) 2016/680.

[202] Article 15(1) Directive (EU) 2016/680.

[203] Article 16(4) Directive (EU) 2016/680.

2016/680. However, in the police and criminal justice context, conflicts pertaining to processing of personal data may arise between the Data Protection Authority and the judicial authorities in order to determine whether a Data Protection Authority may monitor processing done by judicial authorities. In order to limit the discretionary power of the Member States, Directive (EU) 2016/680 provides that the processing of data by judicial authorities must not be affected by its provisions when acting within their judicial capacity. In spite of that, Article 1(3) Directive (EU) 2016/680 permits Member States to maintain a higher level of data protection which may ultimately be a cause of problems.

10.4.3.10 International Data Transfers

Directive (EU) 2016/680 provides rules for international data transfers in its Chapter V.

10.4.3.11 Data Transfers Among Member States

Where personal data are to be transmitted or made available from another Member State, Article 35 (1) Directive (EU) 2016/680 requires five enumerated conditions to be met including that the other Member State has to give its prior authorisation to the transfer in accordance with its national law.[204] However, according to Article 35(2) Directive (EU) 2016/680, Member States shall provide for data transfers without prior authorisation by the another Member State to be permitted if, and only if, the data transfer is necessary for the prevention of an immediate and serious threat to public security of a Member State or a third country, and the prior authorisation cannot be obtained in good time. In these scenarios, the second sentence of Article 35 (2) Directive (EU) 2016/680 requires that the authority which is responsible for giving prior authorisation has to be informed without delay.

10.4.3.12 Data Transfers to Third Countries

With regard to the transfer of personal data to third countries or international organisations, Article 36 (1) Directive (EU) 2016/680 requires that personal information be allowed to be transmitted by a Member State to a third country only if the Commission has decided that the recipient ensures an "adequate" level of protection. The concept of adequate level

[204] Article 35(1)(c) Directive (EU) 2016/680.

of protection has been defined by the CJEU in the *Schrems* case[205] as requiring the third country in fact to ensure, by reason of its domestic law or its international commitments, a level of protection of fundamental rights and freedoms that is essentially equivalent to that guaranteed within the EU. The CJEU has also stated that the European Commission's discretion as to the adequacy of the level of protection ensured by a third Country should be limited, considering, *first*, the important role played by the protection of personal data in the light of the fundamental right to respect for private life and, *secondly*, the large number of persons whose fundamental rights are liable to be infringed where personal data is transferred to a third country without ensuring an adequate level of protection.[206]

In that respect, it should be underlined that data processing in the police and criminal justice context is up until now a field left outside Union law. Practically all Member States have bilateral agreements with third countries permitting the exchange of personal data for law enforcement-related purposes, notwithstanding any "adequacy" finding in respect of the recipients' data protection safeguards. Therefore, here again Directive (EU) 2016/680 had to maintain a careful balance between, on the one hand, the requirements of police and criminal justice work and existing bilateral agreements and, on the other, the requirement for an increased level of personal data protection.

Directive (EU) 2016/680 appears to do little to affect bilateral agreements which are already in place. As a consequence, Directive (EU) 2016/680 automatically turns all bilateral agreements into definite term ones which are in need of amendment to match its standards as soon as the first opportunity arises. However, if Member States – that are called upon, but not obliged to actively seek to amend bilateral agreements in the foreseeable future[207] – do not take action, the prolonged existence of those bilateral agreements which apply lower standards than Directive (EU) 2016/680 could undermine the whole international data transfer edifice.

[205] CJEU, decision of 6 October 2015 in case *Schrems*, C-362/14, CRi 2016, p. 25 at para. 73.

[206] CJEU, decision of 6 October 2015 in case *Schrems*, C-362/14, CRi 2016, p. 26 at para. 78; CJEU, decision of 8 April 2014 in case *Digital Rights Ireland*, C-293/12 and C-592/12, at para. 47 and 48.

[207] See Article 40 Directive (EU) 2016/680.

10.4.3.13 Profiling

The regulation of profiling in Directive (EU) 2016/680 deserves a separate mention. Profiling is especially problematic in the police and criminal justice context because if profiles are misused they can lead to stressful situations for individuals who could be put under surveillance or arrested on the grounds of automated processing of personal data. The compatibility with the presumption of innocence[208] may be questioned.

In this context, it is necessary to underline that Directive (EU) 2016/680 provides substantial and procedural safeguards. According to Article 11(1) Directive (EU) 2016/680, Member States are prohibited from providing for a decision based solely on automated processing, including profiling, which produces an adverse legal effect concerning the data subject or significantly affects him or her, unless authorised by Union or Member State law to which the controller is subject and which provides appropriate safeguards for the rights and freedoms of the data subject. Article 11(3) Directive (EU) 2016/680 also stresses that profiling resulting in discrimination against natural persons shall be prohibited.

References

FATF, International Standards on Combating Money Laundering and the Financing of Terrorism & Proliferation, updated October 2016 (FATF, Paris, 2012), Available at: www.fatf-gafi.org/recommendations.html

M. Gercke, Impact of the Lisbon treaty on fighting cybercrime in the EU. CRI **11**, 75–80 (2010)

S. Peers, EU criminal law and the treaty of Lisbon. Eur. Law Rev. **33**, 507 (2008)

B.R. Roth, State Sovereignty, International Legality, and Moral disagreement, 2005, p. 1, Available at: https://www.yumpu.com/en/document/view/4246351/state-sovereignty-international-legality-and-moral-disagreement

F. Zeder, EU-minimum rules in substantive penal law: What will be new with the Lisbon Treaty? in ERA-Academy of European Law (ed.), *ERA Forum*, vol. 9 (2008), Vienna, pp. 209–27

[208] Article 48 Charter of Fundamental Rights.

PART 3

Case Studies

CHAPTER 11

Case Study: Dark Web Markets

Yara Abdel Samad

11.1 Case Study 1: Hansa

Hansa – ranked once as the largest Dark Web market (seen Chap. 4) in Europe and the world's third largest Dark Web market – was taken over and subsequently taken down by the Dutch National Police after vigorous coordination and cooperation with American and German law enforcement agencies and Europol's support. Hansa, operating on a hidden service of the Tor network, had approx. 3600 dealers with over 24,000 listings of drug products and a smaller scale of businesses in counterfeit documents and fraud tools (Greenberg 2018).

The takeover of Hansa was the result of months of planning, proactive actions and coordination led by the Dutch National Police. In 2016, Europol provided the Netherlands National High Tech Crime Unit (NHTCU) with an investigation lead into Hansa (Europol 2017). This was assisted by a well-established Internet security provider known as Bitdefender, which advised Europol's European Cybercrime Centre (EC3). The lead consisted of a development server found in a data centre

Y. Abdel Samad (✉)
CENTRIC, Sheffield Hallam University, Sheffield, UK
e-mail: Y.Abdel-Samad@shu.ac.uk

B. Akhgar et al. (eds.), *Dark Web Investigation*,
Security Informatics and Law Enforcement,
https://doi.org/10.1007/978-3-030-55343-2_11

of a web-hosting firm in the Netherlands. In fact, it was a test version for new features prior to going live that handled thousands of visits daily (Greenberg 2018).

The NHTCU contacted the web-hosting company and was granted access to the data centre hosting the server. The Dutch police then installed wiretaps on the server to monitor the traffic and watch all packets coming in and out. They discovered that the development server was communicating frequently with a Tor-protected server at the same location on which Hansa's site was running. The server was also communicating with two other servers in a data centre in Germany. The Dutch police was able to make hard drive copies of the entire hard drives of Hansa's servers both in development and production. This included the records of all transactions performed in Hansa's history as well as the conversations that occurred via an anonymized messaging system. This was accomplished without causing service interruptions or network outages on the live site (Darknet Diaries 2018).

Hansa's administrators remained unknown until the Dutch Police found old chat logs on IRC (Internet Relay Chat), an old messaging protocol. The chat logs consisted mainly of information related to the site's activities, such as maintenance and updates and the resolution of disputes among users. Fortunately, they also contained a small number of Bitcoin addresses as well as both administrators' full names and the home address of one of the men in Germany. In fact, both administrators resided in Germany; one was a 31-year-old man who lived in Cologne, while the other was a 30-year-old man in Siegen. The Dutch Police contacted the German authorities and were informed that these two administrators were already being monitored and under investigation for creating an online site for trading pirated e-books and audio books (Darknet Diaries 2018).

The Dutch and the German authorities set up a joint plan to use the German e-books piracy investigation as a cover to take over Hansa without the knowledge of administrators, moderators or users. Unfortunately, the plan encountered a hiccup – as Hansa's server went offline. The administrators discovered that their hard drives had been copied and hence moved Hansa to another location, which was Tor protected, and subsequently distributed once more across anonymized machines globally.

The Dutch and German authorities decided to take their time and continue with their original plan: they agreed not to arrest Hasan's administrators to allow them to gather more evidence to locate the site's servers, so not to depend solely on clues from the administrators' computers. Both authorities spend several months to gather additional evidence trying to locate the new servers.

April 2017 saw a breakthrough as Hansa's administrators made a payment using bitcoin from an address that was in the IRC chat logs discovered earlier by the police. Using the blockchain analysis software Chainalysis, the police found that the payment went to a bitcoin payment provider that operates in the Netherlands. Accordingly, the Dutch police requested further information from the bitcoin payment provider and managed to identify the recipient of the transaction. It was another hosting company in Lithuania (Greenberg 2018).

The cooperation was extended to include the Lithuanian authorities who assisted in tracking down the location of Hansa's server. Simultaneously, cooperation was extended toward US authorities as the FBI alerted the Dutch authorities that they found who was behind AlphaBay (the world's biggest dark marketplace at that time) as well as the location of its servers in Canada. *Operation Bayonet* was commissioned. This operation was a multinational enforcement operation that targeted AlphaBay and Hansa Dark Web markets simultaneously. The goal was to takedown AlphaBay, take control of Hansa and absorb users flooding from AlphaBay to monitor a considerable share of the Dark Web economy and shake users' trust in Dark Web markets, while ensuring the control and surveillance of law enforcement agencies.

Following the plan and making use of the mutual Dutch and Lithuanian legal assistance treaty, the Dutch police sent two agents to the data centre hosting Hansa's server in Lithuania. At the same time, the German Police conducted raids within the homes of Hansa's two administrators, arresting both individuals and seizing their computers while the hard drives were unencrypted. Subsequently, the German police signalled the Dutch police, who immediately began migrating Hansa's data to new servers in the Netherlands that were under full police control (Greenberg 2018). This was possible since, after being caught, the two administrators provided all credentials and passwords needed to access the different parts of the site. Hansa had four moderators who did not know that a takeover had occurred (Darknet Diaries 2018). Sellers and buyers were still able to access Hansa without realizing that the police in the Netherlands and the public prosecution service had seized control of the site (Netherlands Police 2017).

The Dutch Police became in full control of Hansa and turned it into an enormous 'surveillance hub' collecting information and evidence against dealers using the site. The site code was rewritten to log all users' passwords in a decrypted format to save and reuse later in other dark market sites. The police also managed to secretly log the full text of users' messages before being encrypted allowing the capture of home addresses of buyers sent to sellers. Additionally, the police altered the functionality that automatically removed an image's metadata uploaded to the platform, which enabled pulling geolocation data from several photos that sellers took of illegal items. Police also deleted previous photo database enticing vendors to reupload photos to capture the metadata (Greenberg 2018). The Dutch Police further deceived Hansa users to download a homing beacon claiming that the file was a backup encryption key that allowed users to access their bitcoins if the site went down. Once downloaded and opened, the file would run a script pinging back to a police server unmasking the user's IP address giving authorities further clues on where users are located (Darknet Diaries 2018).

During this period, the Dutch police continued acting as the site's administrators, giving support to users, handling complaints and responding to moderators while trade figures grew considerably. On average, 1000 orders were placed daily responding to around 40,000 advertisements. Earlier, Hansa had 1765 different sellers and since control of the site was seized by authorities more than 50,000 transactions took place involving mostly soft drugs and hard drugs. Furthermore, tens of thousands of non-encrypted messages between buyers and sellers were intercepted by the police who managed to identify the delivery address for many orders (Netherlands Police 2017).

Meanwhile, the FBI leading the AlphaBay case was ready to act. The FBI had tracked down the location of the servers in Montreal, Canada, as well as Alexandre Cazès, the owner of Alphabay, who was living in Thailand. The FBI coordinated with the Canadian and Thai authorities to raid the data centre and Alexandre's house at the same time and arrest Alexandre while he was logged into his computer – to have a proof that he was AlphaBay's admin. The mission was successful. Alexandre was arrested in his villa in Thailand and the servers in Montreal were seized and taken offline immediately. When AlphaBay was seized, it had more than 400,000 registered users and 250,000 active listings (Darknet Diaries 2018).

Hansa then became the natural destination for AlphaBay users: more than 5000 users daily moved to Hansa, increasing the usual registration rate eight times. The NHTCU had to shut down new registrations for 10 days, as the Dutch law demands the police to track and report all transaction that took place while the site was under their control to Europol. This included approximately 1000 illegal transactions daily, making the paperwork nearly unmanageable. During this time, only opioid Fentanyl was banned by the Dutch police, being the most dangerous drug. Other drugs on Hansa were traded freely (Greenberg 2018).

After 27 days and around 27,000 illegal transactions later, the NHTCU decided to unplug Hansa on July the 20th and replaced the platform with a link to NHTCU's site. In addition to this, they displayed a list of those involved with drug dealing and purchasing on the Dark Web alongside the message: "We trace people who are active at Dark Markets and offer illicit goods or services. Are you one of them? Then you have our attention" (Greenberg 2018).

During these 27 days, the Dutch Police obtained data of almost 42,000 users, including the foreign addresses of approximately 10,000 buyers, which were handed to Europol to distribute to police agencies across Europe and globally. Also, more than 500 Dutch delivery addresses were reported to couriers and postal services with the intention of stopping the deliveries (Netherlands Police 2017). Furthermore, a dozen of Hansa's top vendors were arrested, and 1200 bitcoins were seized from Hansa, as the NHTCU disabled bitcoin's multisignature transaction functionality. The Dutch police also performed around 50 "knock-and-talks", i.e. visits to the homes of buyers to inform them that they had been recognised through their drug purchases over the Dark Web (Greenberg 2018).

As dark markets may continue to grow and survive, the international cooperation between LEAs such as the Dutch police, the FBI, Europol and the authorities in Germany and Lithuania in the Hansa case is essential to paralyze such illegal activities and damage the credibility and reliability of Dark Web market places. As a law enforcement approach, making use of the combined technical and operational strengths of several agencies proved to be a great success and a clear demonstration of how the collective power law enforcement communities around the world can impede and dismantle serious criminal activities (Europol 2017).

11.2 Case Study 2: The Wall Street Market

The Wall Street Market (WSM), launched in 2016, was one of the largest Dark Web-based markets globally following closely behind Dream. It was taken down in May 2019 as the result of intense cooperation between EU and US law enforcement agencies. When taken down, the WSM had more than 63,000 sales offers placed, over 1,150,000 users and 5400 vendors

(Europol 2019). WSM users significantly increased after the shutdown of AlphaBay and Hansa in 2017 and the shutdown of Dream in April 2019. Furthermore, seizing Silkkitie (Valhalla), one of the oldest darknet marketplace established in 2013, earlier in the same year made Finnish narcotics traders move to WSM increasing further its numbers of vendors and overall users.

WSM had it all – it offered interfaces in six languages including English, French, German, Italian, Portuguese and Spanish. It had many separate categories for products, including drugs, jewellery, equipment and support for credit card fraud and software and malware, among others (The Guardian 2019).

Like most other darknet sites, WSM was accessed through the encrypted Tor-network (see Chap. 4) to shield customers from detection and cryptocurrencies bitcoin and Monero were used in transactions (The Guardian 2019). Administrators were compensated by receiving commission payments that varied between 2% and 6% of the total value of illegal sales facilitated by the site (Europol 2019).

Although investigations into WSM had been ongoing since 2017, they were about to fail due to an exit scam by WSM administrators on the 23rd of April 2019. The WSM administrators suddenly removed all the cryptocurrency stored under their authority. It is believed that alleged owners could have gained around $11 million, if it was possible for them to convert their cryptocurrencies (Coldewey 2019). Buyers and sellers responded angrily about the "Sorry guys we are currently redesigning WSM" message posted by administrators on the 26th of April, as well as a message saying that "maintenance" would last a week (Vaas 2019).

Conversations on Dread, a dark net discussion forum, explained that WSM may have scammed people out of the enormous amount of $30 million worth of cryptocurrencies. This frightened several users who experienced similar exit scams previously, a serious issue with centralized Dark Web marketplaces (Redman 2019).

After WSM's message, one of its moderators who used the moniker Med3l1n and was probably excluded from the exit scam started blackmailing buyers and sellers, requesting 0.05 Bitcoin (~$280) in payment; otherwise, he would provide authorities information about those who had shared, by mistake, their unencrypted details. A few days later, the same moderator leaked the IP address as well as login credentials for the WSM backend located in the Netherlands on Dread. In addition to exposing

WSM server's physical location, this meant that logging into the administrative section of WSM and gaining access to the necessary data to strip the anonymity of sellers, buyers and orders active on the market became possible. As a consequence, the WSM site showed errors to users 6 days after the exposure (Vaas 2019).

The Scam urged investigators in Germany, Europol and the USA to condense efforts and take action to gather and observe additional evidences before administrators could run away and launder the virtual goods. The WSM was created and operated by three administrators from Germany, Klaus-Martin Frost, Jonathan Kalla and Tibo Lousee, who law enforcement agencies managed to trace, identify and capture. The three men were known to US, Dutch and German investigators by the monikers "coder420," "Kronos" and "TheOne" (Reuters 2019). The international investigation revealed that WSM administrators operated another Dark Web marketplace in the past, which was based in Germany and was shut down in 2016. Investigations also linked administrators to computer servers that used to operate WSM and process cryptocurrency transactions, both in Netherlands and Germany (The United States Department of Justice 2019).

One of the WSM's administrators used mainly two VPN service providers to access the WSM infrastructure. Occasionally, one VPN provider connection would cease, exposing the true IP address of the administrator as he continued to access the WSM infrastructure. The same administrator used a UMTS-stick, a dongle for mobile Internet access, registered to a suspected fictitious name. The German federal police (Bundeskriminalamt – BKA) conducted several surveillance operations to identify this UMTS-stick electronically. The surveillance team found that between 5th and 7th of February 2019, this UMTS-stick was used in Kleve, Nordrhine-Westphalia (Germany) at an IT company where Tibo Lousse previously worked as a computer programmer. Shortly, the authorities found Lousee in possession of the UMTS-stick (Coldewey 2019).

The second administrator Kalla, although he used strong VPN, was identified because of meta-data. An IP address assigned to Kalla's home accessed the second VPN provider approximately at the same time as the second VPN provider accessed "administrator's only components" of the WSM server infrastructure. The account linked to the IP address was registered in Kalla's mother's name; during investigations, Kalla admitted he was the user in question (Coldewey 2019).

Identifying the third administrator Forst was the result of thoughtless cross-contamination between cryptocurrency and cryptographic accounts. The German federal police located a PGP public key in the WSM database for "TheOne", a WSM administrator account, which turned out to be the same as the PGP public key for another moniker "Dudebuy" on Hansa marketplace. A financial (bitcoin) transaction, associated with a virtual currency wallet used by Frost, was linked to "Dudebuy". The bitcoin payment processing company's records showed that "Martin Frost" was the buyer. Furthermore, bitcoin wallets used by Frost were recognised by the US Postal Inspection Service by analysing blockchain transactions as well as the information obtained from the proprietary software (Coldewey 2019).

Lousee's, Kalla's and Frost's arrests were the outcome of a well-organised operation led by the German Police. As part of the house searches that took place, "the police officers seized over €550,000 in cash, cryptocurrencies Bitcoin and Monero in 6-digit amounts, several vehicles and other evidences such as computers and data storage" (Imossi 2019). The three Germans were arrested on the 23rd and the 24th of April 2019. Authorities said that administrators stole around $11 million from the accounts of WSM's users (O'Neill 2019).

On the 2nd of May 2019, the BKA, with supervision of the German Public Prosecutor's office, shut down WSM. The BKA worked in collaboration with the Dutch National Police, Europol, Eurojust and different US government agencies including the Drug Enforcement Administration, Homeland Security Investigations, the FBI, Internal Revenue Service, the US Postal Inspection Service and the US Department of Justice (Europol 2019).

Two of the highest-selling suppliers of narcotics on WSM were arrested in the USA during the investigation held by the Attorney General in Los Angeles. This was "Ladyskywalker" who sold opiates such as fentanyl, hydrocodone and oxycodone and "Platinum45" who dealt in methamphetamine, oxycodone and a combined amphetamine prescription drug marketed as Adderall (The Guardian 2019). A 29-year-old Marcos Paulo De Oliveira-Annibale who resided in Sao Paulo, Brazil, was identified as a WSM moderator and was also charged in a criminal complaint filed in the US District Court in Sacramento, California. His moniker "Med3l1n" was connected by prosecutors to his offline identity as he left behind photos and other clues years ago online (Kerb 2019). Annibale operated also as a dealer and represented WSM on Reddit and forums (Scienceexist 2019).

The success of these investigations was the outcome of an operation coordinated by a number of law enforcement agencies across Europe and the USA supported by Europol. "These two investigations show the importance of law enforcement cooperation at an international level and demonstrate that illegal activity on the Dark Web is not as anonymous as criminals may think", according to Europol's Executive Director, Catherine De Bolle (Europol 2019).

Unfortunately, and as expected, new Dark Web markets have emerged to take the lead after the shutdown of Dream and WSM. The majority of these have only been operational for weeks and were launched when users' trust in Dark Web market started to diminish. This included Cryptonia, Agartha, Empire, Nightmare, Core Market and Yellow Brick Road. Some of these markets will naturally get shut down or experience exit scams sooner or later, but as always, more will rise to take their place (Sedgwick 2019).

References

D. Coldewey, How German and US authorities took down the owners of Darknet drug emporium Wall Street Market. *Tech Crunch.* [Online] (2019, May 3), Available at: https://techcrunch.com/2019/05/03/how-german-and-us-authorities-took-down-the-owners-of-darknet-drug-emporium-wall-street-market/?guccounter=1&guce_referrer_us=aHR0cHM6Ly93d3cuZ29vZ2xlLmNvbS8&guce_referrer_cs=cPCk-1J5jt92zr1xKnNk6Q. Accessed 23 Aug 2019

Darknet Diaries, EP 24: Operation Bayonet. *Darknet Diaries.* [Online] (2018), Available at: https://darknetdiaries.com/episode/24/. Accessed 17 Aug 2019

Europol, Double blow to Dark Web Marketplaces. *EUROPOL* [Online] (2019), Available at: https://www.europol.europa.eu/newsroom/news/double-blow-to-DarkWeb-marketplaces. Accessed 23 Aug 2019

Europol, Massive blow to criminal Dark Web activities after globally coordinated operation. *Europol* [Online] (2017), Available at: https://www.europol.europa.eu/newsroom/news/massive-blow-to-criminal-DarkWeb-activities-after-globally-coordinated-operation. Accessed 17 Aug 2019

A. Greenberg, Operation bayonet: Inside the sting that hijacked an entire Dark Web drug market? *Wired – Security*. [Online] (2018), Available at: https://www.wired.com/story/hansa-dutch-police-sting-operation/. Accessed 17 Aug 2019

T. Imossi, Double blow to Dark Web marketplaces. *Association of British Investigators* [Online] (2019), Available at: https://www.theabi.org.uk/news/double-blow-to-DarkWeb-marketplaces. Accessed 23 Aug 2019

B. Kerb, Feds bust up Dark Web hub Wall Street Market. *Krebsonsecurity*. [Online] (2019), Available at: https://krebsonsecurity.com/2019/05/feds-bust-up-DarkWeb-hub-wall-street-market/. Accessed 23 Aug 2019

Netherlands Police, Underground Hansa Market taken over and shut down. *Netherlands Police: Politie*. [Online] (2017), Available at: https://www.politie.nl/en/news/2017/july/20/underground-hansa-market-taken-over-and-shut-down.html. Accessed 17 Aug 2019

P. O'Neill, One of the largest dark net markets 'of all time' falls to police. *Gizmodo*. [Online] (2019), Available at: https://gizmodo.com/one-of-the-largest-dark-net-markets-of-all-time-falls-1834511568. Accessed 23 Aug 2019

J. Redman, Darknet users allege Wall Street Market exit scammed, Possibly Snatching $30M. *Bitcoin* (2019), Available at: https://news.bitcoin.com/darknet-users-allege-wall-street-market-exit-scammed-possibly-snatching-30m/. Accessed 23 Aug 2019

Reuters, Accused operators of illicit 'darknet' market arrested in Germany, Brazil. *Reuters* [Online] (2019), Available at: https://www.reuters.com/article/us-germany-security-darknet/accused-operators-of-illicit-darknet-market-arrested-in-germany-brazil-idUSKCN1S923R. Accessed 23 Aug 2019

Scienceexist, How German and US authorities took down the owners of Darknet drug emporium Wall Street Market – TechCrunch. *Scienceexist*. [Online] (2019), Available at: https://www.sciencetells.co.uk/how-german-and-us-authorities-took-down-the-owners-of-darknet-drug-emporium-wall-street-market-techcrunch/. Accessed 23 Aug 2019

K. Sedgwick, The Darknet rises with 6 new markets. *Bitcoin*. [Online] (2019), Available at: https://news.bitcoin.com/the-darknet-rises-with-6-new-markets/. Accessed 23 Aug 2019

The Guardian, German police shut down one of the world's biggest Dark Web sites. *The Guardian*. [Online] (2019), Available at: https://www.theguardian.com/world/2019/may/03/german-police-close-down-darkWeb-marketplace. Accessed 23 Aug 2019

The United States Department of Justice, Three Germans who allegedly operated Dark Web Marketplace with over 1 million users face U.S. narcotics and money

laundering charges. *The United States Department of Justice* [Online] (2019), Available at: https://www.justice.gov/opa/pr/three-germans-who-allegedly-operated-DarkWeb-marketplace-over-1-million-users-face-us. Accessed 23 Aug 2019
L. Vaas, Dark web marketplace Wall Street Market busted by international police. N*aked Security by SOPHOS*. [Online] (2019), Available at: https://nakedsecurity.sophos.com/2019/05/07/DarkWeb-marketplace-wall-street-market-busted-by-international-police/. Accessed 23 Aug 2019

CHAPTER 12

Case Studies: Child Sexual Exploitation

Alice Raven, Babak Akhgar, and Yara Abdel Samad

12.1 Case Study 1: PlayPen

PlayPen, which was launched in August 2014, only needed a few months to become the most horrendous child pornography dark web site globally before its shutdown in February 2015. PlayPen was an anticipated result of a tremendous increase in child pornography: Located in the USA, the National Centre for Missing and Exploited Children (NCMEC) estimated that over 26 million sexual-abuse videos, and images were watched in 2015, and over 10,500 victims shown in child pornography have been recognised and located by law enforcement agencies between 2002 and 2015 (FBI 2017a).

The dark web has been, and still is, an ideal place for illegal activities including child pornography. According to a study conducted by Spitters et al. (2014), the same year PlayPen was created, they found that a large proportion of the Tor platforms they analysed exhibited illegal or at least controversial content. Adult content (or indicators of such) was identified

A. Raven (✉) • B. Akhgar • Y. Abdel Samad
CENTRIC, Sheffield Hallam University, Sheffield, UK
e-mail: a.raven@shu.ac.uk; Y.Abdel-Samad@shu.ac.uk

B. Akhgar et al. (eds.), *Dark Web Investigation*,
Security Informatics and Law Enforcement,
https://doi.org/10.1007/978-3-030-55343-2_12

on 17% of the platforms in their collection, and half of the content was classified as child pornography.

Shutting down PlayPen, and arresting service beneficiaries in various countries afterwards, was the result of a fruitful cooperation between law enforcement agencies. In December 2014, four months after PlayPen was created, a foreign law enforcement agency (FLA) informed the Federal Bureau of Investigation (FBI) concerning PlayPen's unique IP address having links to a location in the USA. This was the tip that led to the arrest of PlayPen's administrators, allowing it to be taken over it for nearly 2 weeks and then taken down afterwards. Although the FBI already knew about PlayPen and started their investigations, not much was accomplished due to the encrypted nature of Tor's services until the IP address was discovered.

After additional investigations, the FBI was granted a warrant to search the address and as a result confiscated the server, which hosted PlayPen, and obtained a copy of its contents. The site was found on four hard drives on a server in North Carolina (Malik 2018). Steven W. Chase, 58, of Naples, Florida, identified as the main curator and lead administrator of PlayPen, was arrested on the 19th of February 2015 following an intense search of his home which was approved by the court. The forensic examination of seized devices and a computer pursuant to the search revealed that Chase owned thousands of child sexual abuse and exploitation images that included infants and toddlers (The United States Department of Justice 2017).

The foreign law enforcement agency, acting independently and according to its own national laws, seized another child pornography site and operated it 4 months after the seizure. The FLA used a hacking technique of its own to identify at least one user of that site, which led the FBI to identify a suspected moderator of PlayPen. The FLA obtained an IP address and a session identifier to link the IP address to the activity of a specific user account, one of the moderators of that site, by sending this specific moderator a link to a live streamed video that ran on an external platform. The IP address was then provided to the FBI and led to the arrest of David Lynn Browning of Kentucky in July 2015. He was a moderator of the child pornography site seized by the FLA and a suspect of being a moderator on Playpen, according to communications provided by the FLA to the FBI in April 2015. The FLA also obtained the IP address for Michael Fluckiger, another suspected moderator on the seized site and administrator on PlayPen who was arrested in March 2015 (Cox 2016a).

When the FBI took over PlayPen, the site had around 214,000 members, 117,000 postings and an untold number of illicit pictures, videos, and links to additional illegal content (Cox 2016a, b). Images and videos exchanged through the website were classified according to the age and gender of victims and the type of sexual activity. In addition to depending on Tor's anonymity capability, PlayPen members used other advanced means, such as elaborate file encryption to impede law enforcement agencies' efforts (The United States Department of Justice 2017).

The seized copy of PlayPen was put on a server owned by the government in the East District of Virginia. The site was operated by the FBI for 13 days after obtaining a warrant to apply a Network Investigative Technique (NIT) that would assist in revealing the identities of the original users of the site (Electronic Frontier Foundation 2019). NIT is "a term used exclusively by the US government to refer to the methods or tools it uses to access computers of individuals that have taken steps to obscure or mask certain identifying information, like an IP address" (Electronic Frontier Foundation 2019). In fact, NIT in the PlayPen case was a type of malware which was distributed through Tor-hidden services. This malware was designed to penetrate the anonymity granted by the Tor-network by "exploiting a vulnerability in the Firefox web browser, running as part of the Tor Browser, to place computer code on users' computers that would transmit private information back to a law enforcement server outside of the Tor network" (Electronic Frontier Foundation 2019).

Using NITs was not new in 2014. NITs have different forms and have been used by the US government since 2002, if not earlier. Malware has been delivered, using phishing e-mails, to bomb threat suspects. Also, the FBI, as part of "Operation Torpedo" in 2011, inserted an NIT on servers of three different hidden services that hosted child pornography. The NIT used a flash application to ping users' real IP addresses back to a server controlled by the FBI, instead of protecting users' identities by routing traffic through the Tor-network (Cox 2016a, b).

After seizing PlayPen, investigative leads were issued by the FBI to all offices throughout the USA. As a result of the investigation, according to the FBI in May 2017, a large amount of individuals were arrested and prosecuted including 350 individuals in the USA (arrested), 25 users that produced child sexual exploitation material in the USA (prosecuted), 51 US abusers that performed sexual abuse to children (prosecuted) and 55 children in the USA, subjected to sexual abuse, and were successfully identified or rescued. Furthermore, 296 sexually abused children were identified or rescued internationally, and 548 international arrests took place (FBI 2017b). Figure 12.1 includes the figures as announced by the FBI as of May 4, 2017.

The EUROPOL-based European Cybercrime Centre (EC3) was the coordinator and international lead. EC3 received and distributed the information through its network of EU member states as well as the FBI Legal Attaché network (FBI 2017b). This cooperation was part of "Operation Pacifier". In collaboration with the FBI and US Department of Justice, Operation Pacifier was created in January 2015, assisted by Europol and other relevant law enforcement agencies globally to search for and identify the thousands of members of PlayPen. Europol's role was to crosscheck, analyse and complete the data received in order to identify offenders located mainly in Europe. To complement the effort, intelligence packages were prepared and disseminated to law enforcement agencies in various countries such as France, Ireland, Italy, Slovakia, Spain,

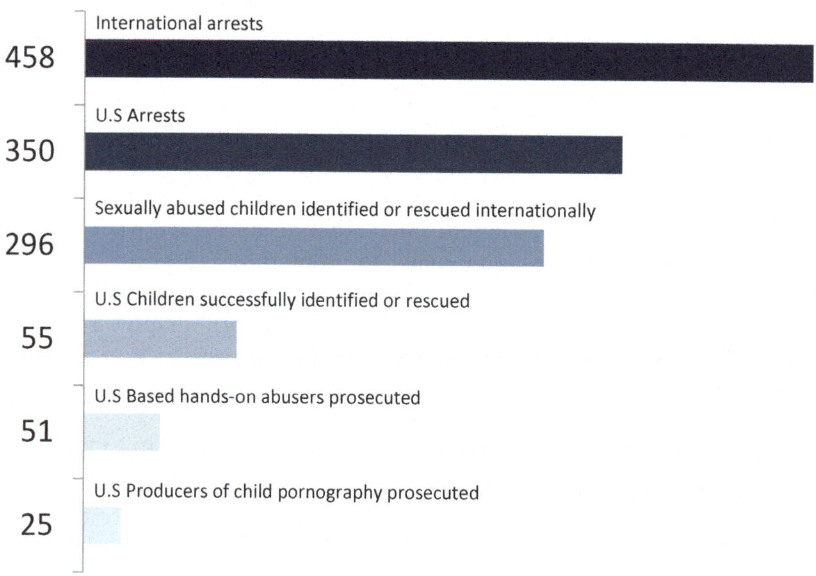

Fig. 12.1 Reproduction of the results of PlayPen investigations as of May 4, 2017 (FBI 2017a)

Switzerland, the UK, Colombia, Croatia and Czech Republic (Europol 2017).

The PlayPen case raised various debates in the media as well as at court, as the FBI decided to continue operating PlayPen for almost 2 weeks before making the decision to shut the platform down, permitting the download of thousands of images of child pornography. While the deputy director of the Department of Justice (DOJ) Keith Becker, who was concerned with the New York-based Child Exploitation and Obscenity Section, argued that allowing the site to remain active was an "investigative necessity" (Carter 2016), the US District Judge Robert Bryan challenged the DOJ prosecutors over their insistence that the government's actions in controlling and secretly operating PlayPen for 2 weeks in 2015 was "innocent", emphasizing that he has "ethical and legal concerns over the Department of Justice's decision to take control of a child-pornography bulletin board and allow the distribution of as many as 1 million illegal images while agents hacked the computers of the site's visitors" (Carter 2016).

Moreover, some prosecutions made in the USA have been challenged considering the legal aspects of the FBI's warrant as well as the prosecutors' rejection to disclose the NIT operations to the defendants. Conversely, several courts supported "governmental actions in dangerous decisions that, if ultimately upheld, threaten to undermine individuals' constitutional privacy protections in their home computers" (Electronic Frontier Foundation 2019). Also some privacy advocates considered that the method of hacking into 8700 devices in 120 countries solely from the authorisation of a single FBI search warrant is unethical, emphasizing that the US Law enforcement agency in question has unlawfully extended its surveillance parameters without the permission of the countries where the computers targeted by the US NIT were hosted and questioning whether it was possible for any foreign intelligence agency to conduct a similar hacking operation that would compromise US citizens' computers. On the other side, US attorneys believed the FBI agents followed the right procedures in getting the warrant and that there had not been other means to capture PlayPen's criminals (Paganini 2017).

In May 2017, Steven W. Chase's trial was held in a federal courtroom in North Carolina where he was sentenced to 30 years in prison, being convicted on various child-exploitation and child-pornography charges. This sentence stemmed from an earlier September conviction for "one count of engaging in a child exploitation enterprise, one count of advertising child pornography, three counts of transportation of child pornography, and one count of possession of child pornography" (Luperon 2017). The jury also returned a special verdict determining that "Chase should be ordered to forfeit all property derived from, involved in, or traceable to his criminal activities, to include his Naples residence" (The United States Department of Justice 2017). The two co-defendants 46-year-old Michael Fluckiger of Indiana and 47-year-old David Browning of Kentucky also received 20-year prison sentences for helping in running PlayPen (Luperon 2017).

12.2 Case Study 2: Matthew Falder

The Matthew Falder case is a clear demonstration of how child sexual abuse material (CSAM) is produced, shared and viewed on the dark web by offenders through the use of sophisticated methods. The case of Matthew Falder from the UK presents the first member of the dark web community "Hurt2theCore" (H2TC) to be arrested and sentenced within the country, which represented a major breakthrough in dark web forum investigations (Nash 2019). This reflects a growing surge in law enforcement agencies analysing cases within dark web forums and recognises how such cases can unravel to reveal often a global scale of transnational CSAM communities that interact on such forums similar to H2TC. The case also showcases from an operational perspective how international cooperation between law enforcement agencies is paramount to the investigation of high-priority cases on the dark web.

Between 2013 and 2017, an intensive investigation led by the Federal Bureau of Investigations (FBI) was conducted to tackle the growing CSAM content that was being distributed through the dark web. This spike in offenders transitioning to dark web platforms was due to the attractive features the dark web platforms offered such as anonymity as key factors (see Chap. 1) to ensuring the content disseminated was concealed and did not reveal the identity of the members accessing the content (Shillito 2019). To achieve this, they developed and hosted a number of websites on the dark web servers, which allowed them to have full transparency of which users were accessing the site and to identify which actors were key distributors of CSAM content (BBC 2018a).

This covert investigative method used is particularly effective when observing dark web platforms, to prevent dark web users becoming aware of the presence of LEAs which could drive them toward more encrypted sites. Furthermore, traditional methods of shutting down illegal sites are proving less effective, as the sites will reappear in other areas of the dark web. Therefore, by hosting dark web websites to track information flow, rather than preventing CSAM on a short-term basis through conventional methods, law enforcement agencies can gain an in-depth and ethnographic understanding of the data collected, to achieve more long-term solutions such as identifying key actors or repeat victims who can then be prioritised (Martellozzo 2015).

Through the discussed method, the FBI gained access to a renowned paedophile site labelled H2TC which is a community that is directly linked to CSAM. The forum is a sub-genre of a CSAM platform known as "hurt-core" that discusses the abuse of children at extreme cases, which in 2013 had 326 accounts created and 160 posts a day (Vice 2015). HT2C was described by the Minister for Security and Economic Crime in the UK as dedicated to discussions and exchanging of content related to rape, murder, sadism, torture, paedophilia, blackmail, and humiliation (HC Deb, Vol. 650, Column 587). From extensively analysing H2TC, a user profile labelled "inthegarden" was identified by the FBI as a highly active user who posted and discussed CSAM and sexually abusive content on multiple accounts surrounding the blackmail of a teenage girl (BBC 2018b). The information posted was traced to a UK-based server, which demonstrates the transnational capabilities of offenders alongside the global scale of CSAM and online dark web forums and communities. As a priority, the details of the case were shared with the National Crime Agency (NCA) within the UK (BBC 2018a).

In parallel to the FBI investigations, the NCA was conducting an investigation into a user labelled "666devil" who posted CSAM of their "daughter" at regular intervals onto dark web forums (Grafton-Green 2019). From the identification of "666devil", the NCA conducted further investigations into the profile which also had an image of the same "daughter" as their profile image and found that the profile discussed abusing and torturing their "daughter" during a period labelled "hell week" which encouraged discussions within dark web forums by asking for requests and advice from other users within the community (BBC, 2018a). From combining the information discovered in the investigation and the information provided by the FBI, the NCA concluded that "666devil" held substantial links to "inthegarden" and a third account labelled "evilmind". The NCA as a result prioritised the case as a major safeguarding investigation, as the profiles displayed severe content discussing CSAM. As shown with the case, the information gathered by the FBI and shared with the NCA became a key identifier of multiple accounts used by one individual within dark web forums. This displays the crucial need for law enforcement agencies to adopt a collaborative framework that shares information regarding CSAM cases between law enforcement agencies on a transnational scale in order to meet the growing globalisation of CSA communities and links between offenders and victims.

To approach the investigation, the NCA created a joint global taskforce between key members tackling CSAM: the UK Government Communications Headquarters (GCHQ), the US Homeland Security, Europol and the Australian Federal Police (Domdouzis 2019). Their investigations revealed the offender used different accounts such as "666devil", "inthegarden" and "evilmind" to approach victims, which later totalled 70 online identities (Grafton-Green 2019). To attract young females, in particular, the offender created a number of false identities using different personas such as taking the role as a female artist named "Liz Candell", "Jess" and "Shona" (BBC 2018a). These accounts were used to advertise posts on the sales site Gumtree, which offered audiences money for artistically posed (often nude) photographs that would be choreographed and directed by the offender to be transformed into "drawings". Other advertisements posted by the offender included dog walkers and babysitters. The offender built upon a number of interactions created from young people interested in the adverts to build a rapport, which differed depending upon the role the offender was adopting. For example, for the female artist the offender mainly discussed that "she" was depressed, because she was unable to have children and enjoyed drawing young people to combat these problems (Grafton-Green 2019).

Once the relationship had been established, the offender facilitated the transition of the conversation from surface platforms such as Gumtree to more encrypted sites such as WhatsApp, where the policing of conversations was limited and the images could be sent by the victim without recognition from any agencies (Halliday 2017). The multiple accounts identified in the investigations proved difficult to uncover due to the methods used by the offender to conceal their digital footprint. They heavily relied on encryption, which can be used to hide messages from external viewers in order to spread and store CSAM. When on the encrypted sites, the offender manipulated victims into revealing intimate details about themselves including their sexual history and to produce and share sexual imagery (Grafton-Green 2019). Once the offender received the images, they used the content as leverage to blackmail the victims further into sending increasingly severe CSAM images including "eating their own faeces, tampons and drinking urine", in return for the CSAM remaining private (Nash 2019).

The CSAM was collated and traded by the offender onto dark web communities such as H2TC for approval by the community members. The appraisal of the offenders on H2TC was established through a points system, which increased if members were consistently active on the site and uploaded content. In this case, the offender was promoted by the community point's infrastructure to "VIP Rapist" status (BBC 2018a). By achieving different levels of status, members of H2TC were able to view more content and become more immersed into the sites as a result, which entrenches the difficulty for offenders to disband themselves from the site. Incorporating gaming strategies into dark web communities is not an uncommon phenomenon; competitive and hierarchal characteristics are used as a unique selling point for the sites which encourages the dissemination of CSAM content and active contribution by members that are crucial for the sustainability and development of the platforms (Vice 2015). This influences members to complete tasks and move upwards in the hierarchy which reinforces their commitment to the platforms. These communities when joined are difficult to disband from, which can leave offenders in compromising positions in which they have to prove their worth by collating and sharing CSAM. It is these sites which escalated the offences committed by Falder. If law enforcement agencies could in essence diffuse the gaming tactics used by these sites, this could substantially decrease the audience's interest and immersion within the platforms.

Two years into the joint investigation into the three profiles, a 16-year-old girl reported that a "man disguised as a female artist" contacted her on Gumtree after she posted a babysitting advert online, asking her to send a naked image to be used for artwork in return for £1200. The girl accused the account user of being male rather than an artist, and in response the user sent pornographic images as a form of "revenge". The girl reported the account to the NCA, which led to a crucial development in the investigation of "inthegarden" and the corresponding accounts that allowed the agency to discover a further 50 victims whose information confirmed that the three profiles under investigation derived to one account (Halliday 2017). The suspect's webmail accounts were accessed on the surface web, which allowed the agencies to identify the victim ("daughter") labelled in the profile pictures. The account was linked to an address in Birmingham that was owned by Matthew Falder, who was later identified as the offender.

To ensure that sufficient evidence was collated that proved and clearly showed that Falder was the offender, the UK law enforcement agencies and NCA conducted a 3-month covert investigation to record his Internet use by filming him using his laptop on a train in the UK, for example. The outcome of the investigation revealed the connection between the reported cases to the history of dark web posts associated with multiple usernames (BBC 2018a). In June 2017, the investigation was successful in collating sufficient evidence. Therefore, Falder was arrested by undercover authorities while working at the University of Birmingham. They immediately seized two of his devices at his workspace and raided his home to recover digital evidence which revealed the seismic scale of Falders offences which began in 2009 (BBC 2018b).

At the beginning, before his offences became more serious, the investigation found that Falder recorded his friends, peers and unrelated individuals through hidden cameras in the bathrooms and bedrooms of his family homes, college dorms and personal home (Halliday 2017). These recordings were stored and used to blackmail victims to send compromising images of similar nature to stop the images from being disseminated (Grafton-Green 2019). It was Falder's transition onto dark web communities such as H2TC which facilitated the development of the crimes into more intense and harmful methods in order to reach the approval and rise of status within the community.

In terms of hardware encryption, the methods used by Falder highlight the nuanced methods offenders are using alongside the manipulation of advanced technologies to conceal CSAM from law enforcement agencies. Due to his profession as a professor at Birmingham University, Falder had a large knowledge of technology, computers and the dark web which allowed him to avoid detection for his offences for the duration of 4 years (Nash 2019). When his computer was seized, the police discovered a USB stick which was double-encrypted to conceal the contents. From this the police worked in collaboration with computer experts to uncover the material on the USB. Once uncovered, the USB revealed a file marked "BM" which was presumed to be an acronym for "blackmail". Within the file, they uncovered 484 compromising images of victims that were collated by Falder throughout his offending (BBC 2018a). They also uncovered a "paedophile manual" developed by Falder which was created to inform other dark web forum users on H2TC about tips on how to become a successful paedophile which included methods to befriend children (Grafton-Green 2019). This was accompanied by diverse information provided by dark web forums by offenders to share "best practices" such as how to remove meta-data from images so that they cannot be associated with the offender by law enforcement agencies (Vice 2015). It is this information shared on dark web communities such as H2TC which reinforces the activity and continuity of the groups. It also reflects on how offenders can entice and encourage each other to commit CSAM to gain a reputation within the community.

Following 3 days of questioning, Falder admitted ownership of the three accounts "evilmind", "666devil" and "inthegarden". Falder was taken to court and presented guilty to 137 of the 188 offences outlined, with 45 of the 300 victims from a number of countries including England, the USA and Australia speaking within the courtroom. The number of victims identified from the case reveals the sheer scale of CSAM crimes and how the globalisation of technology allows offenders to reach new parameters and communities of vulnerable young people. He was sentenced to 32 years in prison with 6 years additionally on licence (Nash 2019).

From analysing cases such as CSAM, which have become such prevalent crimes where content is widely disseminated on the dark web, law enforcement agencies need to develop new methods of intercepting and preventing these communities from expanding and encouraging offenders to commit crimes through nuanced methods such as points systems. The use of the dark web has made these crimes extremely difficult to detect and police, particularly in terms of identifying users or closing down sites that promote the activity (Domdouzis 2019). The Matthew Falder case provides an interesting case study that reveals a number of effective methods used by law enforcement agencies to meet the rising standards in offenders' methods and technologies (Grafton-Green 2019). The joint cooperation between several agencies throughout the investigation, in particular, provided crucial information that aided the development of the case through the sharing of content discovered and accounts identified. Furthermore, the case reveals how agencies such as the FBI used innovative methods on the dark web to identify the initial profiles Falder used on H2TC. It is these methods that law enforcement agencies need to develop further to ensure that the dark web is effectively analysed, and cases are prioritised efficiently.

12.3 Case Study 3: Operation Sweetie

The growth of CSAM-related cases increases the pressure for law enforcement agencies to find innovative methods of tackling new technologies used by offenders (Martellozzo 2015). A particular strand of CSA, which emerges on the dark web, is "webcam sex tourism". This method involves live exchanges between offenders and victims who are in most cases children, including children performing sexual acts through webcams (BBC 2013). This development has been exacerbated by organised crime groups and in some instances families capitalising on this offence, by recruiting children and forcing them to communicate with offenders in enclosed spaces for financial gain (Terre des Hommes 2019). The scale of the problem was outlined by the United Nations (UN) and FBI who estimated that 750,000 users globally are online for webcam sex tourism with children on over 40,000 public chat rooms and dark web sites (Broadhurst 2019). This case study presents an innovative covert operation conducted in the Netherlands that aimed to tackle this challenge by identifying offenders on dark web platforms using advanced technologies such as artificial intelligence.

The globalisation of the Internet together with developments in user security including encryption such as on the dark web allows offenders to use the Internet as part of online communities which benefit from the "flexibility, scalability and survivability" of such platforms (Martellozzo 2015). Sites particularly within the dark web can receive high levels of activity on a daily basis, some up to 500 page views per second (BBC 2013). To address the challenges of growing anonymity online, law enforcement agencies are developing and exploring new covert techniques, in particular, to identify and arrest offenders. This case study reflects a shift from reactive to proactive, intelligence-led law enforcement agency approaches, of which operation such as "Operation Sweetie" is an example (Noorlander 1999).

Operation Sweetie is a key case study for understanding how LEAs are developing new processes to reinforce a proactive approach to combatting the rise in webcam sex tourism. The 10 week sting operation was carried out by the nongovernmental organisation (NGO) Terre des Hommes, an International children's rights organisation in Switzerland that advocates for the importance of tackling webcam sex tourism (Martellozzo 2015). In 2013, Operation Sweetie was created alongside the animation company "Lemz", which involved the development of Artificial Intelligence (AI) technologies to create an avatar which was designed to reflect a 10-year-old Philippine girl labelled "Sweetie_1000" (Sweetie) (Terre des Hommes 2019). The operators created user profiles with usernames such as "10 f Philippines" which reveals the age, gender and location of the child reflecting authentic profiles used by children in regular cases (Santbrink and Guijt 2013).

The use of AI is an innovative method of re-enacting how webcam sex tourism works, presenting a new way of tackling the offence. Sweetie was used by the operators at Terre des Hommes to enter chat rooms and forums on the dark web. It should be noted here that, to ensure best practices were kept, the operators did not actively approach any users on the sites and would only communicate with offenders who created the initial contact with Sweetie (Broadhurst 2019). The presentation of the profile – including the avatars age in the username and reinforcing the age of Sweetie as 10 year old – was crucial for the operation to ensure that (on record) the offenders acknowledged that Sweetie was underage but still accepted this fact (Santbrink and Guijt 2013). This ensured that the cases adhered to ethical guidelines and did not involve entrapment which induces offenders to commit crimes they otherwise may not have committed. This also prevented the risk of any potential "blue on blue" cases which could hinder the operation.

Once approached by a user, Sweetie would offer the offender a "preview" which would be free and include a visual of Sweetie which reassured offenders that they were talking to a young girl (Santbrink and Guijt 2013). This differs largely from original conversations between LEAs and offenders, as they would not be able to provide this certification through previews, which could deter paranoid offenders from providing personal information and consequently hinder the investigation. For example, when the researchers at Terre des Hommes attempted to conduct a similar method without using Sweetie, the offenders often distrusted the account and as a result held back sharing personal information (Santbrink and Guijt 2013).

Sweetie has provided law enforcement with a solution to overcome this obstacle by giving previews to offenders that were reluctant to show their faces without confirmation. Sweetie instantly gained a level of trust. Following the preview, offenders would request to begin webcam conversations with her in which the operators would control how she behaved, spoke and her movements through motion capture and fictional pre-texts (Acar 2017). The operation of Sweetie was conducted by two operators working within Terre des Hommes. Moving into the further stages of the conversations, the offenders would be asked if they were willing to transfer $20 to an account "associated" with Sweetie to reinforce the fact that Sweetie was performing sexual acts for money rather than for pleasure and from there would provide their Skype address so that the conversation could progress into webcam sex (Acar 2017).

Once the Skype address of the offender was captured, the details were gathered by the operators and the conversation was terminated. The information concerning the offender was given to a further two operatives within the institution who would conduct an open source investigation (OSINT) into the details of the individual. By conducting OSINT, the operators were able to retrieve key information concerning the offender's name, location, related images and any other relevant information that could be used as an identifier (Martellozzo 2015). The operation involved a joint action and co-collaboration approach with law enforcement agencies (LEAs), which meant that any information collated was shared with the relevant LEAs in order to effectively improve the investigative process into webcam sex tourism (Martellozzo 2015).

The results of Operation Sweetie proved to be highly successful and highlighted the severity of the crime and the overwhelming number of cases for investigation. Within the duration of the 10 weeks operation, over 20,172 offenders contacted Sweetie, and 1000 users offered money to continue the conversation on Skype (Santbrink and Guijt 2013). This reveals how just for one child profile, there are a colossal number of potential offenders who actively want to engage in webcam sex tourism (Terre des Hommes 2019). In particular, the investigation revealed the seismic transnational scale of the offence, as the names identified through OSINT were located in Britain (110), the USA (254) and India (103), as a number of examples (Terre des Hommes 2019). The demographic nature of the offender profiles shows how technological advancements combined with globalisation facilitates the growth of offences on dark web platforms that can allow offenders to communicate with potential victims throughout the dark web communities, which are often undetected from more traditional policing methods.

Operation Sweetie demonstrates how new approaches can be developed and exercised to unravel these dark web communications and lead to the identification of offenders. Impressively, in several countries in which information was shared by Terre des Hommes, a number of arrests occurred. In 2019, a British national Matthew B., Norwegian national

Jomar K. and Dutch national Hans v.C. were arrested as a result of Operation Sweetie (BBC 2018b). These cases show how innovative methods to identify offenders involved in webcam sex tourism are successful for combatting the challenges the dark web presents.

The operation proved that proactive approaches can be effective means to tackling the growing use of the dark web to conduct CSA. The results of the operation, in particular, helped to raise awareness of webcam sex tourism in the public which subsequently caused a global outrage and widened the traditional understandings of CSA offences to incorporate Internet-facilitated offences. The growth in dark web methods of webcam sex tourism not only created a societal impact but also became a high priority within the United Nations. This led to a number of vulnerable children at risk specifically in the Philippines being safeguarded from becoming victimised (Terre des Hommes 2019). Understanding the dark web methods and demographics of offenders such as in Operation Sweetie shows how legal, societal, academic and expert knowledge understandings of the crimes can be implemented into new preventative strategies.

Although the operation was conducted as a small-scale project, scholars have discussed the possibility for similar methods for LEAs to encourage the use of more proactive approaches. The technique of using Artificial Intelligence to model children as avatars can be used on a global scale to identify offenders and prevent children from being involved in the sex webcam industry. This will also remove ethical and legal strain from current profiles used by LEAs on dark web platforms that are easily exposed, when operators are not being able to present themselves as a child and gain rapport with offenders. Traditional methods such as these (creating fake profiles on the dark web) will only reveal a small scale of the problem, as more sophisticated offenders will be deterred from interacting with profiles which appear to be controlled by LEAs. The ability for operators to use avatars to convince offenders that the profile is legitimate removes current constraints and allows operators to operate as a neutral actor (Acar 2017). This new approach reflects a proactive, effective and ethically implemented method of challenging the growth in CSA cases to directly prevent the rise in offending, identify the methods employed by offenders including platforms used to increase LEAs' knowledge base and most importantly safeguard vulnerable children approached by offenders through the dark web.

References

K. Acar, Webcam child prostitution: An exploration of current and futuristic methods of detection. Int. J. Cyber Criminol. **11**(1), 98–109 (2017)

BBC, Computer-generated 'Sweetie' catches online predators. *BBC News*. [Online] (2013), Available at: https://www.bbc.co.uk/news/uk-24818769. Accessed 19 Aug 2019

BBC, Dark web paedophile Matthew Falder jailed for 32 years. *BBC News*. [Online] (2018a), Available at: https://www.bbc.co.uk/news/uk-england-43114471. Accessed 15 Aug 2019

BBC, Dark web paedophile Matthew Falder's sentence reduced. *BBC News*. [Online] (2018b), Available at: https://www.bbc.co.uk/news/uk-england-45875275. Accessed 17 Aug 2019

R. Broadhurst, Child sexual abuse images and exploitation materials, in *Cybercrime: The Human Factor*, ed. by R. Leukfeldt, T. Holt, (Routledge, Oxon, 2019)

M. Carter, Judge has 'ethical and legal' concerns over FBI running a massive 'dark web' child-porn site. *The Seattle Times* [Online] (2016), Available at: https://www.seattletimes.com/seattle-news/crime/judge-has-ethical-and-legal-concerns-over-fbis-massive-child-porn-sting/. Accessed 20 Aug 2019

J. Cox, How the FBI located suspected admins of the dark web's largest child porn site. *Vice* [Online] (2016a), Available at: https://www.vice.com/en_us/article/jpgm7d/how-the-fbi-identified-suspects-behind-the-dark-webs-largest-child-porn-site-playpen. Accessed 20 Aug 2019

J. Cox, The FBI's 'Unprecedented' hacking campaign targeted over a thousand computers. *Vice* [Online] (2016b), Available at: https://www.vice.com/en_us/article/gv5ggq/defense-lawyers-claim-fbi-peddled-child-porn-in-dark-web-sting. Accessed 20 Aug 2019

K. Domdouzis, Detecting psycho-anomolies on the world-wide web: Current tools and challenges, in *Advances in Psychology Research*, ed. by D. Konstantinos, (NOVA Science Publishers, Hauppauge, 2019)

Electronic Frontier Foundation, The playpen cases: frequently asked questions. *EFF* [Online] (2019), Available at: https://www.eff.org/pages/playpen-cases-frequently-asked-questions#whatisanit. Accessed 30 Oct 2019

Europol, Major online sexual abuse: operation leads to 368 arrests in Europe. *Europol*. [Online] (2017), Available at: https://www.europol.europa.eu/newsroom/news/major-online-child-sexual-abuse-operation-leads-to-368-arrests-in-europe. Accessed 20 Aug 2019

FBI, The source of child pornography – working to stop the sexual exploitation of children. *FBI*. [Online] (2017a), Available at: https://www.fbi.gov/news/stories/the-scourge-of-child-pornography. Accessed 20 Aug 2019

FBI, Playpen creator sentences to 30 years. *FBI.* [Online] (2017b), Available at: https://www.fbi.gov/news/stories/playpen-creator-sentenced-to-30-years. Accessed 20 Aug 2019

P. Grafton-Green, Who is Matthew Falder? The Birmingham University researcher jailed after admitting 137 charges including encouraging child rape. *Evening Standard.* [Online] (2019), Available at: https://www.standard.co.uk/news/crime/who-is-matthew-falder-the-birmingham-university-researcher-jailed-after-admitting-137-charges-a3770506.html. Accessed 19 Aug 2019

J. Halliday, Cambridge graduate admits 137 online sexual abuse crimes. *The Guardian* [Online] (2017), Available at: https://www.theguardian.com/uk-news/2017/oct/16/cambridge-graduate-pleads-guilty-to-137-online-sex-abuse-crimes. Accessed 19 Aug 2019

A. Luperon, Creator of world's largest child porn site gets 30-year prison sentence. *Law & Crime* [Online] (2017), Available at: https://lawandcrime.com/high-profile/creator-of-worlds-largest-child-porn-site-gets-30-year-prison-sentence/. Accessed 20 Aug 2019

N. Malik, Terror in the dark: how terrorists use encryption, the darknet, and cryptocurrencies. *The Henry Jackson Society.* [Online] (2018), Available at: https://henryjacksonsociety.org/wp-content/uploads/2018/04/Terror-in-the-Dark.pdf. Accessed 20 Aug 2019

E. Martellozzo, Policing online child sexual abuse: The British experience. Eur. J. Policy Stud. **3**(1), 32–52 (2015)

M. Nash, MAPPA: sex offenders and managing 'the other' in the community, in *Multi-Agency Working in Criminal Justice*, ed. by A. Pycroft, D. Gough, (Policy Press, Bristol, 2019)

P. Noorlander, The impact of the human rights act 1998 on covert policing: Principles and practice. Int. J. Human Rights **3**(4), 49–66 (1999)

P. Paganini, Privacy groups claim FBI hacking operation in the PlayPen case was unconstitutional, *Security Affairs*, 11 February 2017., https://securityaffairs.co/wordpress/56181/laws-and-regulations/privacy-groups-playpen-case.html. Accessed 20 Aug 2019

M. Shillito, Untangling the 'dark web': An emerging technological challenge for the criminal law. Inf. Commun. Technol. Law **28**(2), 186–207 (2019)

M. Spitters, S. Verbruggen, M. Staalduinen, Toward a comprehensive insight into the thematic organization of Tor hidden services, in *Proceedings of the 2014 IEEE Joint Intelligence and Security Informatics Conference*, 2014, pp. 220–223. https://doi.org/10.1109/JISIC.2014.40

Terre des Hommes, Sweetie 2.0: Stop webcam child sex, *Terre de Hommes* [website], https://www.terredeshommes.nl/en/programmes/sweetie-20-stop-webcam-child-sex. Accessed 20 Aug 2019

The United States Department of Justice, Florida Man Sentenced to Prison for Engaging in a Child Exploitation Enterprise, *The United States Department of*

Justice [website], (1 May 2017), https://www.justice.gov/opa/pr/florida-man-sentenced-prison-engaging-child-exploitation-enterprise. Accessed 20 Aug 2019

A. Van Santbrink, H. Guijt, Webcam child sex tourism, *Terre des Hommes* (2013), https://www.terredeshommes.nl/sites/tdh/files/uploads/research_report.pdf. Accessed 21 Aug 2019

Vice, You wanted darker web?, *All Things Vice,* September 11 2015., https://allthingsvice.com/2015/09/11/you-wanted-darker-web/#more-740. Accessed 19 Aug 2019

CHAPTER 13

Case Study: Match Fixing on the Dark Web

Alice Raven

13.1 Case Study 1: Match Fixing on the Dark Web

The process of betting within the sporting field is not a new phenomenon; it reveals an international industry which attracts a huge audience of users. For this reason, it is no surprise that there is a parallel and more sinister market which capitalises upon sporting events through "match fixing". This form of crime involves better using unorthodox methods in line with a network of individuals involved within teams to ensure that the bet that they make comes through. Due to the high pay-outs for both the better and those on the inside of the teams, this approach has revolutionised the way sporting offences need to be addressed. Market fixing has become a phenomenon that is not solely limited to football, but recognisable throughout all forms of sport, for instance, basketball as highlighted by the infamous case of the NBA referee Tim Donaghy manipulating scores for the Gambino crime family (Andrews 2016). The association of match fixing with organise crime groups (OCGs) has become a point of interest,

A. Raven (✉)
CENTRIC, Sheffield Hallam University, Sheffield, UK
e-mail: a.raven@shu.ac.uk

B. Akhgar et al. (eds.), *Dark Web Investigation*,
Security Informatics and Law Enforcement,
https://doi.org/10.1007/978-3-030-55343-2_13

due to the billions of pounds the betting industry accumulates during large-scale events (Ellen and Hancock 2018).

The international scale of these crimes has been identified in a number of studies. In Europe alone, match fixing is notable to staggering amount: The ICSS (2017) found that 34.7% of athletes believed a game they were playing in was fixed and 12.6% reported to be involved in a fixed game (ICSS 2017). This is followed by 15% of players reporting they have been approached to fix a match (in 2018–2019), which shows how this crime has become prominent as a "natural" behaviour that comes hand in hand with sporting events (ICSS 2017).

The most vulnerable sports players approached are young players beginning within a team or older players that are reaching the end of their employment. Organised crime groups recruit and manipulate young players that are due to enter teams or are progressing in their sport. By recruiting these players, OCGs are able to confirm the inclusion of young people and ease them into the grooming process, so that when they reach their peak, they are fully immersed into the offending. Furthermore, once a young player has been recruited for one match fix, they can be blackmailed by the OCG to continue offending or otherwise jeopardise their career. Older and more experienced players are particularly vulnerable when they are reaching the end of their careers. Many players within this position have overspent their earnings or would like to increase their income before retirement, for instance. Therefore, they become a source of income for the OCGs, who often lose the sense of risk in offending due to their financial situations believing that once they retire they will not be exposed to offending (Andrews 2016). By recruiting these two types of vulnerable players, OCGs can ensure that their businesses are reliable and gain the most income in order to expand and grow.

The main platform used by OCGs, networks and individual "offender entrepreneurs" to coordinate, advertise and communicate all matters of match fixing is the Dark Web. As a modern technique, offenders are now exploiting the Dark Web to advertise their services to potential gamblers. This is largely due to the levels of anonymity and perceived safety from law enforcement available to users on the Dark Web (see Chaps. 1 and 2). As recognised by Ellen and Hancock (2018), many of these sites are well-established and are designed in a way that signals their "legitimacy" to the audience such as including logos and in some cases switching between the dark and open web. These sites tend to cover a range of sports including football, tennis and ice hockey (Ellen and Hancock 2018). It is this form of legitimacy which ensures that the administrators receive the largest influx of clients, for example, by shadowing legal betting company advertising and marketing strategies they are able to duplicate the market on the Dark Web.

The financial structure of these institutions is facilitated by the Dark Web, particularly due to the rise in online currencies such as Bitcoin, Monero, Ethereum and Ripple which disassociate user identities from the transactions made (Ellen and Hancock 2018). The prices of the services provided by the sites varies along a number of factors including the seriousness of the game, the odds at hand and the level of expertise of the players. For example, games such as finals or tournaments would have a larger cost in comparison to a regular game, which can increase depending on the odds due to the higher pay-outs for betters. Prices may vary from $99 for inside information regarding one game which could benefit the better's odds to $1500 for 2–3 guaranteed matches that are fixed, or as reported by Andrews can be 0.3BTC which equivalates to $200 for a small game (Ellen and Hancock 2018). Andrews (2016) on the more extreme end discusses a site labelled "BET FIXED MATCH", which claimed to have a global network of 58 athletes, referees and sports agents, charged betters $25,000–800,000 per game. These examples show that the financing is a dynamic sphere that differs between each site.

To explore how these sites worked, in particular, with respect to finances, Andrews conducted a 2-week interview over the Dark Web with an offender called "Frank" who was an established match fixer. Frank operated using an "onion" URL which is operated on a Tor browser, using Appaloosa Chat to run the site which mainly focused on Football league matches in America. To provide legitimacy checks for new customers, Frank would provide them with a "free trial" in which he gave information about a low-key game. An example of a game provided to Andrews was a score for the Audax Rio EC U20 versus Olaria RJ match, which correctly ended 7–0 as Frank predicted. Frank would receive the scores from an interconnected network of experts through an email using ECSDA encryption, a "cryptographic algorithm". Using these secure methods provided by the Dark Web allows offenders to share sensitive information without being uncovered, which also maintains the anonymity of the networks involved. In terms of finances, Frank allows betters to use two options: the first is via direct bank transfer using Bitcoin and the second by using a "middleman" on the Dark Web who holds the better's money in storage until the score of the game has been confirmed and is correct according to the information provided (Andrews 2016).

In the following, we discuss match fixing investigations conducted on the Dark Web in football, using a particular case in 2013 that involved a Europol investigation into links between match fixing within a number of European football matches and an Asian-based OCG.

13.2 Case Study 2: Investigating Match Fixing in Football

By far the most prominent sport that is associated with match fixing is football. The process of match fixing can include traditional methods such as fixing the score, or more abstract methods such as "spot fixing" which involves a wider range of bets, for example, the number of throw-ins in a match or the timing of penalties (Europol 2013). As mentioned previously, large sporting events are targeted by OCGs on the Dark Web due to the high levels of interest by betters and their potential for large amounts of revenue. An example of an investigation conducted by FIFA for the 2018 World Cup in Russia reveals the scale of these offences during popular events and showcases the preventative methods being used to prevent match fixing. A Saudi Arabian referee Fahad Al-Mirdasi was banned by FIFA from refereeing at the World Cup due, as he was suspected of being involved in match fixing on the Dark Web (Ellen and Hancock 2018).

A more prominent investigation called "Operation VETO" presents a pivotal moment in law enforcement agencies' efforts into combatting match fixing and the innovative methods used by offenders, being labelled as the largest football match fixing investigation to be conducted within Europe. This case displays the effectiveness of LEAs working cooperatively across Europe to improve the tackling of match fixing offences, as they become more globalised in nature. From 2011 to 2013, Europol alongside 13 law enforcement agencies from European countries (including Germany, Finland, Hungary, Austria and Slovenia) together with Eurojust and Interpol conducted an 18-month investigation labelled "Operation VETO", into a number of suspected fixed football matches being facilitated on the Dark Web (Europol 2013).

The investigation targeted a large network of 400 individuals within the football field which included match officials, club officials, players and members of organised crime groups (Ellen and Hancock 2018). This investigation emphasises the increasing number of OCGs turning to match fixing on the Dark Web for financial gain. The football matches under investigation included the World Cup and European Championship qualifiers, two Champions League Ties and several top matches in European Leagues (Europol 2013). This highlights the increase in interest surrounding large matches due to their larger audiences, higher number of betters and better financial opportunities; World Cups, in particular, see billions of euros in potential revenue from gambling (Ellen and Hancock 2018). The investigation included the analysis of 13,000 emails alongside intelligence reports from Europol (Europol 2013).

The results of the investigation led to the identification of a global network of 425 individuals located in 15 countries that were identified in match fixing attempts over 380 professional level matches (Ellen and Hancock 2018). This was a well-established organised crime group based in Singapore that coordinated the offending, believed to be involved in a large proportion (150) of matches under suspicion of being fixed. The OCG used high-end bribes such as $100,000 per match to manipulate and target players and other experts within the field that were vulnerable. The income of the Asian OCG was substantial and reflects its popularity on the Dark Web. Its customers spent a total of 16 million euros from which the OCG gained 8 million euro profit (Ellen and Hancock 2018).

Europol identified that the group had been operating on the Dark Web on Asian markets, who's OCGs created a transnational organisation with European facilitators. This case also identified the OCG's cooperation with groups in Russia and other syndicates, which implies a growth in OCGs working together to increase the parameters of match fixing. For example, Europol suggests that in one fixed match over 10 countries with up to 50 offenders can be involved, representing a much wider framework of offences on a globalised scale (Europol 2013). The results of Operation VETO led to a large number of arrests, including 14 arrests in Germany with sentences totalling up to 39 years in prison (Europol 2013).

The above examples demonstrate that match fixing in Europe has become a transnational market which is largely associated with advanced organised crime groups who are coordinating the offences through innovative methods on the Dark Web. To challenge these nuanced methods, LEAs are employing a collaborative approach to combat match fixing. For example, the EU and Council of Europe alongside Europol have developed the 'KEEP CRIME OUT OF EUROPE' project. This prevention approach has proved particularly effective in advising LEAs in how to prevent match fixing and develop a global understanding of how the offence is gaining prominence in major sporting events (Europol). As technologies progress and the number of OCGs and offenders turning to the Dark Web to conduct match fixing increases, LEAs need to continue to work collaboratively as an International response to produce effective and impactful investigations.

References

J. Andrews. Can you buy fixed soccer match results on the Dark Net? *Vocativ* [website] (2016), https://www.vocativ.com/330391/soccer-match-fixing-dark-net/index.html. Accessed 24 Aug 2019

L. Ellen, J. Hancock. Cyber Threats to the 2018 FIFA World Cup Russia, *Mishcon de Reya* [website] (2018), https://www.mishcon.com/news/publications/cyber-threats-to-the-2018-fifa-world-cup-russia. Accessed 24 Aug 2019

ICSS. 35% Athletes believe matches at their level were fixed, says forthcoming study. *ICSS* [website] (2017), http://theicss.org/2017/07/23/35-athletes-believe-matches-at-their-level-were-fixed-says-forthcoming-study/. Accessed 25 Aug 2019

Europol, Results from the largest football match-fixing investigation in Europe, *Europol* [website] (2013), www.europol.europa.eu/newsroom/news/update-results-largest-football-match-fixing-investigation-in-europe. Accessed 24 Aug 2019

Index

B. Akhgar et al. (eds.), *Dark Web Investigation*, Security Informatics and Law Enforcement,
https://doi.org/10.1007/978-3-030-55343-2

The manufacturer's authorised representative in the EU is Springer Nature Customer Service Centre GmbH, Europaplatz 3, 69115 Heidelberg, Germany. If you have any concerns regarding our products, please contact ProductSafety@springernature.com

Printed and bound by CPI Group (UK) Ltd, Croydon, CR0 4YY
22/09/2026
02221062-0001